The Governance of Regulators

Driving Performance at Ireland's Environmental Protection Agency

This work is published under the responsibility of the Secretary-General of the OECD. The opinions expressed and arguments employed herein do not necessarily reflect the official views of OECD member countries.

This document, as well as any data and map included herein, are without prejudice to the status of or sovereignty over any territory, to the delimitation of international frontiers and boundaries and to the name of any territory, city or area.

Please cite this publication as:
OECD (2020), *Driving Performance at Ireland's Environmental Protection Agency*, The Governance of Regulators, OECD Publishing, Paris, *https://doi.org/10.1787/009a0785-en*.

ISBN 978-92-64-38648-8 (print)
ISBN 978-92-64-44374-7 (pdf)

The Governance of Regulators
ISSN 2415-1432 (print)
ISSN 2415-1440 (online)

Photo credits: Cover © Leigh Prather - Fotolia.com, © Mr.Vander - Fotolia.com, © magann - Fotolia.com.

Corrigenda to publications may be found on line at: *www.oecd.org/about/publishing/corrigenda.htm*.
© OECD 2020

The use of this work, whether digital or print, is governed by the Terms and Conditions to be found at *http://www.oecd.org/termsandconditions*.

Foreword

Good regulatory outcomes depend on more than well-designed rules and regulations. They also require bodies to administer these rules to ensure that the right policy outcomes are realised. Regulators are at the delivery end of the policy cycle and their job is inherently a complex one, requiring neutral engagement with a variety of actors, including government, citizens and regulated entities.

The model of arms-length regulators, based on strong technical capacity, transparency, autonomy and constructive engagement with stakeholders, can help regulators tackle this complex landscape. However, regulators need to be correctly equipped to carry out these fundamental tasks and stay abreast of evolving contexts. The good governance of regulators helps ensure that regulatory decisions are made on an objective, impartial and consistent basis, without conflict of interest, bias or improper influence.

To support regulators as they face these challenges, the OECD has developed a framework to assess and strengthen their organisational performance and governance structures. The framework analyses regulators' internal and external governance, including their organisational structures, behaviour, accountability, business processes, reporting and performance management, as well as role clarity, relationships, distribution of powers and responsibilities with other government and non-government stakeholders.

This report applies this Performance Assessment Framework to Ireland's Environmental Protection Agency. This is the first time that the OECD has applied the framework to an environmental regulator, rather than an economic regulator, demonstrating its versatility across different contexts and sectors.

The report finds that the EPA is a highly trusted and respected institution. However, the context in which it operates is evolving rapidly. For example, a significant new environmental policy has recently been introduced, and pressures on licensing and enforcement functions are increasing as economic activity in Ireland picks up. These changes demand a well-thought-through strategic response by the organisation about its future role and objectives. This report identifies opportunities for the EPA to build on its strong reputation and continue to ensure its effectiveness as a modern regulator and employer.

This report is part of the OECD work programme on the governance of regulators and regulatory policy, led by the OECD Network of Economic Regulators and the OECD Regulatory Policy Committee, with the support of the Regulatory Policy Division of the OECD Directorate of Public Governance. The Directorate's mission is to help government at all levels design and implement strategic, evidence-based and innovative policies that support sustainable economic and social development.

Acknowledgements

This report was prepared by the OECD Public Governance Directorate (GOV), under the leadership of Marcos Bonturi, Director, and of Nick Malyshev, Head of GOV's Regulatory Policy Division. It was co-ordinated and drafted by Martha Baxter and James Drummond, under the guidance of Anna Pietikainen. Alexis Durand contributed to the draft and Shelly Hsieh carried out initial preparatory work for the review. Substantive comments were provided by Nick Malyshev and Eugene Mazur, Policy Analyst at the Environment Directorate. The work received the encouragement and support of Marcos Bonturi, and Irène Hors, Deputy Director of GOV. Jennifer Stein co-ordinated the editorial process and Andrea Uhrhammer provided editorial support.

Three peer reviewers provided extensive input and feedback throughout the development of the review: Jonas Landstad Fjeldheim, Strategic Advisor to the Director, Norwegian Environment Agency; Ana Barreto Albuquerque, Member of the Executive Board of Portugal's Water and Waste Services Regulation Authority (ERSAR); and Ian Tait, Director of Network Regulation, Water Industry Commission for Scotland (WICS), United Kingdom. Members of the OECD Network of Economic Regulators (NER) discussed a draft of the report during their meeting on 5 November 2019.

The report would not have been possible without the support of the EPA and its staff. The team would in particular like to thank the following colleagues for their assistance in collecting data and information, organising the team's missions to Ireland and providing feedback at different stages of the review: Laura Burke, Director-General; Gerard O'Leary, Deputy Director-General and Director of the Office of Communications and Corporate Services; Eimear Cotter, Director of the Office of Environmental Sustainability; Matt Crowe, Director of the Office of Evidence and Assessment; Micheál Lehane, Director of the Office of Radiological Protection and Environmental Monitoring; Tom Ryan, Director of the Office of Environmental Enforcement; Darragh Page, Programme Manager, Shirley Murphy, OEE Support Team and Michelle Leahy, Office of the Director-General. Interviews and comments from the various EPA Offices, government, industry and civil society stakeholders during the review process contributed to the analysis presented in the report.

Table of contents

Abbreviations and acronyms — 7

Executive summary — 10

Assessment and recommendations — 13
   Role and objectives — 13
   Input — 19
   Process — 24
   Output and outcome — 36
   References — 41

1 Regulatory and sector context — 42
   Environmental policy and regulations in Ireland — 43
   References — 48

2 Governance of Ireland's Environmental Protection Agency — 52
   Role and objectives — 53
   Input — 68
   Process — 75
   Output and outcome — 95
   Notes — 98
   References — 98

Annex A. Methodology — 100

Tables

Table 1. EPA budget by category, real (millions EUR) and percentage of total income	20
Table 2. Extract of the questionnaire on Workforce	23
Table 3. EPA's five offices and their functions	29
Table 4. Responsibilities for licensing/permitting and enforcement at the EPA	31
Table 5. Eight high-level metrics and indicators included in the PDA 2018	38
Table 2.1. Co-ordination with other state and regulatory bodies	60
Table 2.2. EPA's strategic plan 2016-2020	66
Table 2.3. EPA Budget by category, real (EUR millions) and percentage of total income	69
Table 2.4. Staff by category, 2018	71
Table 2.5. EPA Human Resources Development Strategic Framework 2017-2021	72
Table 2.6. Workforce movement at the EPA, 2016-18	73
Table 2.7. Responsibilities for licensing/permitting and enforcement at the EPA	82
Table 2.8. Types and number of licenses issued by the EPA	84

Table 2.9. Overview of indicators for enforcement activities in industrial and waste sectors, 2015-17	84
Table 2.10. Consultative documents, 2014-18	90
Table 2.11. EPA external committees	91
Table 2.12. Appeals	93
Table 2.13. Indicators reported by EPA to DCCAE	97
Table A A.1. Criteria for assessing regulators' own performance framework	102

Figures

Figure 1. EPA organisational structure	28
Figure 2. Types of EPA enforcement actions	33
Figure 3. Compliance and engagement spectrum	36
Figure 1.1. Ireland's public institutions	43
Figure 2.1. EPA's functions	55
Figure 2.2. Linking EPA's vision and mission with its annual work programme	65
Figure 2.3. The five dimensions of independence identified by the Guidance	68
Figure 2.4. EPA organisational structure	78
Figure 2.5. Types of EPA enforcement actions	85
Figure A A.1. The OECD Best Practice Principles on the Governance of Regulators	102
Figure A A.2. Input-process-output-outcome framework for performance indicators	103

Boxes

Box 1. Redesigning the web presence of the Norwegian Environment Agency	15
Box 2. Building constructive relations with the executive, the case of Great Britain's energy regulator Ofgem	19
Box 3. Performance indicators for people management from the Treasury Board of Canada Secretariat	22
Box 4. Business Planning and Strategic Oversight at Canadian Transportation Agency	25
Box 5. Transparency at Mexico's National Hydrocarbons Commissions	27
Box 6. Water quality testing in Portugal	31
Box 7. Principles of responsive regulation applied to promoting compliance	36
Box 8. Measuring organisational and policy performance: the Canada Energy Regulator's departmental results framework (Canada)	38
Box 9. Key Performance Indicators	39
Box 2.1. Legislation	53
Box 2.2. Memoranda of Understanding	59
Box 2.3. Creating a culture of independence	68
Box A A.1. The input-process-output-outcome logic sequence	101

Abbreviations and acronyms

ABP	Planning Board (*An Bord Pleanála*), Ireland
AER	Annual Environmental Report
ARC	Audit and Risk Committee
ARERA	Regulatory Authority for Energy, Networks and Environment (*Autorità di Regolazione per Energia Reti e Ambiente*), Italy
ASN	Nuclear Safety Authority (*Autorité de sûreté nucléaire*), France
BATs	Best Available Techniques
BI	Behavioural insights
C&AG	Comptroller and Auditor General
CAFE	Clean Air for Europe
CCAC	Climate Change Advisory Council
CEO	Chief executive officer
CI	Compliance investigation
CNH	National Hydrocarbons Commission (*Comisión Nacional de Hidrocarburos*), Mexico
CO_2	Carbon dioxide
COP	Conference of the Parties
CRU	Commission for Regulation of Utilities
CSO	Central Statistics Office
CTA	Canadian Transportation Agency
DAFM	Department of Agriculture, Food and the Marine
DBEI	Department of Business, Enterprise and Innovation
DCCAE	Department of Communications, Climate Action and Environment
DHPLG	Department for Housing, Planning and Local Government
DPER	Department for Public Expenditure and Reform
DPP	Director of Public Prosecutions
EEA	European Environment Agency
EC	Executive Committee
EDEN	Environmental Data Exchange Network
EIA	Environmental Impact Assessment
EIAR	Environmental Impact Assessment Report

ELV	End of Life Vehicles
EPA	Environmental Protection Agency
ERC	Executive Risk Committee
ERSAR	Portugal's Water and Waste Services Regulatory Authority
ETS	Emissions Trading Scheme
EU	European Union
EUR	Euros
Euratom	European Atomic Energy Community
FPN	Fixed Payment Notice
FSAI	Food Safety Authority of Ireland
GMOs	Genetically modified organisms
HFCs	Hydrofluorocarbons
HHI	Herfindahl Hirschman Index
HIQA	Health Information and Quality Authority
HR	Human resources
HRD	Human Resources Development
HSA	Health and Safety Authority
HSE	Health Services Executive
IAEA	International Atomic Energy Agency
ICT	Information and Communications Technology
IED	Industrial Emissions Directive
IEN	Irish Environment Network
IFA	Irish Farmers Association
IPC	Integrated Pollution Control
IPCC	Intergovernmental Panel on Climate Change
KPIs	Key performance indicators
LAWPRO	Local Authority Waters Programme
LEMA	Licensing, Enforcement, Monitoring and Assessment
MLN	Management and Leadership Network
MoU	Memorandum of Understanding
NCMC	National Co-ordination & Management Committee
NDC	Nationally determined contribution
NER	Network of Economic Regulators
NGO	Non-governmental organisation
NIECE	Network for Ireland's Environmental Compliance and Enforcement
NIR	Non-ionising radiation
NPDWAG	National Pesticides in Drinking Water Action Group
NRGI	Natural Resource Governance Institute
NTFSO	National Transfrontier Shipment of Waste Office

NTIG	National Technical Implementation Group	
NWCPO	National Waste Collection Permit Office	
NWPC	National Waste Prevention Committee	
OCCS	Office of Communications and Corporate Services	
ODS	Ozone depleting substances	
OECD	Organisation for Economic Co-operation and Development	
OEA	Office of Environmental Assessment	
OEE	Office of Environmental Enforcement	
OES	Office of Environmental Sustainability	
ORM	Office of Radiation Protection and Environmental Monitoring	
PAC	Public Accounts Committee	
PAFER	Performance assessment framework for economic regulators	
PCQA	Water quality testing programmes	
PDA	Performance Delivery Agreement	
PDO	Protected Disclosures Officer	
PMDS	Performance Management and Development System	
POPs	Persistent Organics Pollutions	
RAL	Remedial Action List	
RBMP	River Basin Management Plan	
RPII	Radiological Protection Institute of Ireland	
SEAs	Strategic environmental assessments	
SMN	Senior Management Network	
UNFCCC	United Nations Framework Convention on Climate Change	
VOCs	Volatile organic compounds	
WEEE	Waste Electrical and Electronic Equipment	
WERLAs	Waste Enforcement Regional Lead Authorities	
WFD	Water Framework Directive	
WPAC	Water Policy Advisory Committee	
WWDL	Waste Water Discharge License	

Executive summary

Ireland's Environmental Protection Agency (EPA) was established in 1993 to protect and improve Ireland's environment. Reflecting its reputation for delivering results, the EPA's responsibilities have expanded over time in step with new legislation and EU directives and following the merger with the Radiological Protection Institute of Ireland in 2014. The EPA has built up a reputation as a trusted and respected body that stakeholders recognise for its scientific integrity. Its technical and administrative independence are reinforced by a strong internal culture of independence.

The EPA faces an evolving context. Pressures on licensing and enforcement functions are increasing as economic activity picks up; public scrutiny of environmental issues is growing; and a new whole-of-government Climate Action Plan seeks to fundamentally change the way Ireland tackles the climate emergency. Within this context, the EPA must make important strategic decisions about its future role and objectives, ensure its attractiveness as a modern employer and cutting-edge regulator, and demonstrate the impact of its work through clear performance reporting.

Role and objectives of the EPA

The EPA has an expansive mandate for environmental protection and has been innovative within its regulatory powers to achieve its objectives. The EPA does not set government policy but draws on its substantial evidence and expertise to inform the policy-making process, often relying on the strength of relationships with policy makers rather than on formal structures. A recent change in parent department from the Department of Housing, Planning and Local Government to the Department of Communications, Climate Action and Environment has been disruptive for the EPA and the executive. The EPA is able to co-ordinate with a large number of public bodies and its Network for Ireland's Environmental Compliance and Enforcement is seen as a best practice at the European level. The increasing number of actors in the environmental sphere creates a risk of overlapping mandates; in this context, co-ordination is more important than ever.

Key recommendations

- Define the EPA's role and strategy in the changing policy context. While the responsibility for policy making rests with government, there may be opportunities for the EPA to engage more proactively in policy development and evaluation, leveraging its expertise and ensuring its evidence and research is policy relevant.
- Fulfil the function of a knowledge provider with more effective communications.
- Discuss, agree and align expectations with parent departments.

Input

The EPA is mostly funded by government budget, including an Environment Fund that is used for essential expenses but is designed to diminish over time. The remaining income is earned from licencing fees and enforcement, some of which is not set on a cost-recovery basis. As the EPA does not have full autonomy in the allocation of its resources, managing peaks and troughs in workloads across its different areas of activity is challenging. The EPA has developed an ambitious human resources development (HRD) strategy that seeks to foster innovation and manage change.

Key recommendations

- Strengthen co-ordination processes with senior levels in parent departments and the Department of Public Expenditure and Reform to co-ordinate solutions to budget and human resource related issues.
- Secure the sustainability of financing by advocating for a review of licensing fees, the allocation of essential expenses to more stable income streams, and stronger medium-term budget commitments.
- Ensure the attractiveness of the EPA as an employer with modern HR practices, harnessing the potential of the HRD strategy by attaching measurable targets to its goals. Setting these targets provides an opportunity for the EPA to ensure that skills are fit-for-purpose as the organisation evolves.

Process

A full-time Executive Board manages the EPA, with most Board meetings dedicated to technical decision making. The governance arrangements and wide responsibilities of Board members highlight the importance of reserving sufficient time for discussions on strategy and seeking diverse external perspectives to strengthen decision-making. Some functions and subject areas are split among different EPA offices and locations, creating challenges in terms of efficiency, consistency of approach and messaging.

The EPA demonstrates independence in its regulatory functions. It carries out licensing and permitting through a transparent and detailed process, which could be streamlined further. The new inspections and enforcement strategy focuses on compliance and takes risk into consideration, sometimes using numerical measures that may be susceptible to bias rather than outcome measures to guide decisions. Several stakeholder engagement processes are in place, but early-stage consultation is not systematic.

Key recommendations

- Isolate opportunities to discuss strategic matters and diversify input into decision making by continuing to bring in external perspectives and fresh ideas. Strengthening the links between the Board and its external committees could be one way to achieve this.
- Take stock of the distribution of activities across offices with the goal of bringing together functions (e.g. enforcement) or subject areas (e.g. climate change).
- Fully implement and monitor with performance indicators the new compliance and enforcement policy to adhere to the principles of responsive regulation, focussing more on outcomes.

Output and outcome

Recent innovations in the EPA's data collection, management and analysis processes have made significant efficiency gains for both the organisation and the entities it regulates. Building on its data collection functions, the EPA monitors and reports on Ireland's environment and regulated entities. While the EPA carries out this work effectively, the information is often difficult to find and navigate online. There is limited engagement with regulated entities around performance.

The EPA's strategic plan sets out goals and associated outcomes for the organisation, but the lack of measurable targets means that it cannot easily be used to monitor the organisation's performance. Parallel processes for monitoring with several sets of indicators create an unnecessary burden. Furthermore, metrics focus on outputs rather than outcomes, missing an opportunity to use performance monitoring to drive improvement.

Key recommendations

- Redesign the EPA's web presence to ensure that data and information are easier to find, understand and use.
- Better engage with regulated entities around performance, e.g. organise events to identify best practices and recognise 'champions' as a way to build trust and drive compliance.
- Develop a unified, outcome-based system for performance assessment and reporting linked to the strategic plan; complement this with clear, plain language communications to show the impact of EPA activities to its diverse stakeholders.

Assessment and recommendations

Ireland's Environmental Protection Agency (EPA) was established in 1993 as a public regulatory body with administrative and technical independence to protect and improve Ireland's environment. Over time the EPA's responsibilities have expanded beyond those originally set out in its founding statute – the Environmental Protection Agency Act, 1992 – in step with new regulations, EU directives and following the merger with the Radiological Protection Institute of Ireland in 2014.

The EPA has established itself as a trusted and respected body for environmental and radiological protection that is recognised for its scientific integrity. It has been given responsibility for regulating an increasing number of areas on account of its reputation to deliver and it operates with a strong culture of independence. EPA data and reports are the reference for knowledge on Ireland's environment and the EPA is seen as an authoritative voice on environmental issues. It networks effectively at the European level, where it has gained a reputation as an innovative, open organisation with many good practices to share.

The EPA operates in a challenging space. It is responsible for implementing environmental and radiological legislation from national and EU levels that is complex and fragmented. The EPA is under the aegis of two government departments and has been building a relationship with a new parent department since 2016. It must also co-ordinate with numerous public bodies operating in the field of environmental protection or related areas, with limited formal co-ordination structures.

This is in many ways a pivotal moment for the EPA. Pressures on the EPA's functions are increasing as economic activity in Ireland picks up. At the same time, public interest in and media scrutiny of environmental issues, in particular the response to climate change, are growing. A new whole-of-government Climate Action Plan (June 2019) seeks to fundamentally change the way Ireland tackles the climate emergency.

Within this context, the EPA must make important strategic decisions about its future role and objectives: does it focus on maintaining its identity as a technical agency, producing highly regarded environmental data and policing compliance, or does it want to further evolve towards a more central space, moving beyond compliance to become a lead agency in encouraging better environmental performance across sectors? Addressing this question will be vital to ensure that the EPA remains relevant in a rapidly changing environmental landscape and provides an opportunity to become a driving force in protecting and improving Ireland's natural environment. It also has many implications for the governance of the EPA.

Role and objectives

Mandate and functions

The EPA has an expansive mandate to "protect and improve Ireland's environment" that has evolved organically over the years following changes in environmental policy and legislative

frameworks in Ireland. The EPA's responsibilities have expanded beyond those originally set out in its founding statute in step with new regulations, EU directives and following the merger with the Radiological Protection Institute of Ireland in 2014. Today EPA's licensing, permitting and enforcement activities cover waste, wastewater, industrial emissions (emissions to air, water and land, generation of waste, noise), greenhouse gases, contained use and controlled release of genetically modified organisms (GMOs), sources of ionising radiation, and volatile organic compounds (VOCs). Its monitoring, analysing and reporting functions span a broader range of environmental areas. These include air quality, water quality (rivers, lakes, bathing water, drinking water…), radiation levels, biodiversity, species and habitats (reporting functions only), greenhouse gases, waste generation and management, and land and soil. These changes have taken place incrementally over the years, resulting in a broad mandate.

The EPA reports on the state of the environment from a holistic perspective, yet its regulatory responsibilities primarily concern emissions and discharges, a distinction that is not always clear to all stakeholders. The EPA is entrusted with a wide range of monitoring and reporting responsibilities (fulfilled, for example, by the publication of the State of the Environment report). Its responsibilities for licensing, enforcement and collecting data from/monitoring regulated entities are much narrower, focusing on activities by particular actors in a given area (for example, waste disposal by large operators but not smaller waste operators, which come under the supervision of local authorities). Its regulatory responsibilities do not cover all environmental policy areas. For example, biodiversity is under the remit of the National Parks and Wildlife Service under the Ministry of Culture, Heritage and the Gaeltacht. This creates challenges that the agency must overcome to define and project a clear corporate identity that is understood by all stakeholders.

The EPA strategic plan for 2016-2020 clearly defines the organisation's three overarching functions as "regulation", "knowledge" and "advocacy", but in practice they are fragmented across the EPA in a manner that may hinder consistency. Licensing, inspections and enforcement form the core of the EPA's regulatory activities. Its knowledge functions include monitoring and reporting on a wide range of environmental outcomes and co-ordinating and funding a significant research programme to advance knowledge on environmental protection. Its advocacy work aims to drive more environmentally-friendly behaviour by citizens, consumers and businesses. In practice, these functions are fragmented across teams and locations, with scope to provide a clearer narrative of goals and resources and consolidate approaches. This is illustrated in how the EPA presents its work in corporate documents, which list ten areas of responsibilities but appears to mix functions (e.g. licensing) with sectors (e.g. water).

The EPA has been innovative within its regulatory powers to achieve its institutional objectives. Legislation provides the EPA with enforcement powers to inspect, investigate and prosecute. The EPA mostly relies on criminal law in its sanctioning activities, with some administrative sanction powers in specific areas. The reliance on criminal law means that cases have to be brought to a criminal standard of proof. This is an administratively heavy process for an outcome that often results in a small fine from the courts, as environmental penalties are capped at relatively low levels in legislation. To overcome these constraints, the EPA has developed innovative tools and funded research and programmes aimed at applying behavioural insights (BI) to improve environmental outcomes. This includes multiple programmes under the National Waste Prevention Programme to reduce waste, the National Dialogue on Climate Action, raising awareness about radon, and improving enforcement through the National Priority Sites, which "names and shames" industrial and waste licensed sites that have recorded poor environmental performance, in order to prompt behaviour change.

The EPA does not set government policy but draws on its substantial data, evidence and scientific expertise to inform the policy making process through a number of channels. The EPA makes formal submissions on draft legislation put out for public consultation; presents to committees of parliament on a wide range of topics that then informs debate and discussion. The EPA is often called upon to respond to informal requests from government departments on the development of policy and draft legislation, particularly where technical or scientific input is required. The EPA is a highly respected body that is recognised for its scientific rigour, high quality data and expertise on the environment nationally and at the European level. This expertise is a very valuable contribution to policymakers and stakeholders at large.

The policy context is changing and the EPA needs to take a strategic decision concerning its future direction. Building on its well established position as an authority on the environment, there could be opportunities for the EPA to provide further evidence-based advice to support the national policy making process. In the rapidly developing environmental policy space this is a significant opportunity for the EPA and it is urgent that it proactively defines its identity, particularly at this critical juncture as Ireland operationalises the all-of-government Climate Action Plan released in June 2019. There are potentially important responsibilities stemming from the plan that the EPA is well-placed to lead on. However, if the EPA is unwilling or unable to be closely engaged in the process it risks being overlooked, undermining its relevance. This moment presents an opportunity for the organisation to define its role in the policy space also with regard to other areas of its work such as waste. The EPA Board needs to take a strategic decision on the direction in which it wishes to take the organisation in this changing context.

Currently, advice or input into policy development tends to rely on the strength of relationships with the relevant government policymakers rather on formal or public structures. The EPA Act empowers the EPA to advise the government "of its own volition" on environmental protection and related matters and it appears more could be made of this opportunity without conflicting with the EPA's independence. This function encompasses giving advice to the government on any proposals for legislative change or other policy matters, as well as reporting and making recommendations on particular environmental issues or problems. More generally, there also seems to be demand from other stakeholders for the EPA to be more proactive and vocal on environmental issues, in particular on the response to climate change, while recognising that decision-making responsibility for policy rests with the government

The forthcoming strategic plan for the period from 2021 is an opportunity for the EPA to re-examine its role and objectives in light of these developments. The EPA operates in the framework of five-year strategic plans, the latest of which (*Our Environment, Our Wellbeing* 2016-2020) was developed and revised in an open and collaborative process internally and externally for consultation. The EPA could continue this good practice for the next plan.

The EPA's performance of its fundamental functions could be strengthened by an improved website. The EPA website has the potential to be a vital tool for fulfilling the EPA's core functions as a knowledge provider, environmental advocate and regulator but is widely acknowledged to not be fit-for-purpose in its current form. It contains a large amount of information but is often hard to navigate and many links are broken. Links and documents related to EPA processes and outputs of interest for stakeholders are partly on display, while others, such as calls for public consultation, are located in hard-to-find places (Box 1).

Box 1. Redesigning the web presence of the Norwegian Environment Agency

The Norwegian Environment Agency owned several websites and domains that it had to keep up to date. The large number of sites resulted in numerous problems: confused users; redundant content on several sites; competition between its sites on search engine rankings; a low number of unique visitors across the sites; inconsistent messaging that made its branding less visible; and difficulty in keeping the same content up to date across the different sites.

The agency therefore decided to consolidate its sites and services in order to maintain its content in a much more manageable way that would support its brand. Nine websites and five administrative information systems with editorial content were consolidated into one website with the user task in focus:

The Environment Agency now focuses on optimising individual landing pages for users who use search engines, with less emphasis given to navigation from the home page. The Environment Agency uses web analytics and software services (e.g. Google Analytics, Siteimprove) to help improve the website with content, search engine optimisation, analytics, accessibility and data privacy.

Result

Target audiences

When navigating to the website, the first thing the user sees is three entrance points for different target audiences: private individuals, businesses and authorities. These entrances are task oriented, complete with guides that helps users to solve a specific task.

- Private individuals refers to tasks such as "How to buy a fishing licence";
- Businesses refers to tasks such as "How to apply for a discharge permit";
- Authorities refers to tasks such as "How to manage outdoor recreation areas".

These guides were previously PDF files. On the new site, these huge documents are now interactive step-by-step guides that further help the Environmental Agency's digitisation process. These documents are now easier to maintain and update through the content management system (CMS). An example of a digital step-by-step guide (in Norwegian) is available here: https://www.miljodirektoratet.no/ansvarsomrader/klima/klimakvoter/klimakvoteregisteret/slik-bruker-du-klimakvoteregisteret/

Areas of activity

The website displays eight areas of activity of the Environment Agency. In these eight categories users can find everything that the organisation does and says but that is not a specific task. The categories are based partly on the agency's organisational structure and partly on the results of a large survey that was carried out with people from within and outside the organisation. This ensured that the labels given to each area of activity correspond to what most users think and expect. Getting to know what users expect to find was an important part of the process in which the Environment Agency invested much effort. It also conducted interviews with relevant real-life users for every guide or area that was revised or digitalised, gleaning valuable information each time.

Miljøstatus – "State of the Environment"

There is a module for the State of the Environment in Norway. This is sourced from a subdomain that was previously a stand-alone website managed by the agency. This is one of the nine websites that was consolidated into one. By creating subdomains of previously stand-alone sites, they can be managed through the same CMS, making updating and avoiding redundant content more efficient, simple and manageable.

News and events

Users can also find news and events from the Norwegian Environment Agency on the homepage. While these are important for the agency, the user statistics show that they are less important than the tasks that users come to the website to perform, so the news and events section is located lower down on the front page.

Source: Information provided by the Norwegian Environment Agency, 2019.

Recommendations

- **Define** the EPA's role and strategy in the changing policy context.
  - While the EPA should maintain and build on its strong technical focus and existing deliverables, the improvement in the economy and the introduction of far-reaching environmental policy through the Climate Action Plan requires a well thought through strategic response from the EPA. The organisation must maintain its policy relevance and ensure that the EPA remains a relevant, centrally involved and strong evidence-based voice for the environment in this new policy landscape.
  - The climate action plan is redefining actors' roles and functions. The plan provides a great opportunity to build the EPA role (for example, the consolidation and enhanced utilisation in EPA of inventories and projections).
  - If the decision is made to take a more proactive role in providing input to policy development, ensure this is done in a transparent manner. While the responsibility for policy making ultimately rests with government, there may be an opportunity for the EPA to engage more proactively in policy development and evaluation of policy impact on the ground, leveraging its considerable expertise and networks.
  - The EPA must also ensure that the data and information that they put forward are relevant to the policy debate. Research products should be reviewed to continue to maintain quality but also offer a "fast-track" option for heightened policy relevance.
- **Clearly communicate** the EPA identity of who they are and what they do to the general public, regulated entities, industry associations and civil society groups.
- **Differentiate** between roles and resources targeted on advocacy, advice and guidance. The role of an environmental protection agency is distinct from that of an environmental NGO. If the EPA is to provide evidence-based advice to the government on policy, this should be done in a predictable and transparent manner and respond to a clear expectation from the executive. If its advocacy functions are to focus on consumer and citizen behaviour change, this should be clearly stated, as advocacy is one of three overarching functions in the strategic framework. Guidance is a third distinct area that focuses on supporting regulated entities to achieve compliance with regulations.
- **Fulfil** the function of a knowledge provider with more effective and outward communications. The EPA has identified many of the steps it needs to take in its communications strategy, and should focus on implementing the priority actions therein. Among the most important is a significant overhaul of the website and other communications tools to ensure accessibility to data and information. This could be informed by "user journey" research and behaviourally-informed techniques to understand how and why users access the site, as well as what to make most prominent.

Relations with the executive and institutional co-ordination

A recent change in parent department to the Department of Communications, Climate Action and Environment (DCCAE) and reallocation of functions between government departments has been a disruptive time for both the EPA and the executive. DCCAE has been the EPA's parent department since 2016. Prior to this, the EPA had been under the aegis of the Department of the Environment, Community and Local Government (now re-named the Department for Housing, Planning and Local Government, DHPLG) with which it had developed long-standing relationships. While time may resolve some of these issues, in the immediate term it is important to ensure clarity around expectations on both sides: what DCCAE expects of the EPA, and what the EPA can potentially offer to support DCCAE in its work. The need for clarity around expectations is made more urgent in the context of defining roles and responsibilities in the implementation of the Climate Action Plan.

Further complexity in relations with the executive is introduced through the division of EPA portfolios between two departments. While DCCAE is the EPA's parent department, DHPLG has responsibility for water policy, and so the EPA retains an important relationship with its former parent Department in this policy area. The three parties occasionally meet at a senior level and mechanisms have been put in place to try to clarify and manage relations between the EPA and the two departments: a tripartite Oversight Agreement and associated Performance Delivery Agreement (PDA) is in place and both departments establish key performance indicators (KPIs) for the EPA to monitor its performance. However, the PDA is fairly generic and does not appear to result in open and meaningful dialogue between the parties, mainly reiterating the EPA's legislative obligations and listing the reports it is expected to publish.

The EPA is able to co-ordinate with a large number of public bodies mostly thanks to personal relationships, in the absence of over-arching formal co-ordination frameworks. Some relationships are defined through formal co-ordination mechanisms, such as MOUs or structures introduced through legislation. However, given the size of the administration, personal connections between staff in government departments and public bodies are common and seem to greatly facilitate communication and co-ordination where they exist.

An increasing number of actors are active in the field of environmental protection and policy, creating a risk of overlapping mandates. For example, the recent establishment of the Climate Change Advisory Council (CCAC), whose secretariat is provided by the EPA but which is an independent public body, may create overlapping responsibilities. The new Climate Action Plan will also redefine roles and responsibilities, including that of the CCAC, raising the need for the EPA to be actively involved in discussions around its operationalisation.

The administrative network for co-ordination between the EPA and Ireland's 31 local authorities appears to be operating well, and is seen as an innovative model among European countries. Local authorities have significant environmental protection responsibilities in Ireland, and the EPA has a supervisory role under Section 63 of the EPA Act 1992 and at the same time has a statutory role to provide advice and assistance. Since 2004, the EPA and local authorities have operated the Network for Ireland's Environmental Compliance and Enforcement (NIECE), which provides the framework for both oversight and support. The network seems to be working well and is an example of good practice. There nevertheless remains an inherent tension in the relationship between the EPA and local authorities due to the EPA's dual role as regulator of local authorities and provider of advice and assistance.

The EPA co-ordinates effectively at the European level and is a highly respected partner. The EPA Director-General is the current Chair of the European Environment Agency (EEA) Management Board, and the EPA is active in EEA activities at a technical level. The EPA actively participates and networks effectively in other European fora, for example, in expert groups of the European Commission.

Recommendations

- **Discuss, agree and align** expectations and relations with parent departments. The EPA needs to proactively rebuild bridges following the reform. Clarity is needed on all sides about the parents departments' expectations of the EPA and what the EPA can potentially offer to support the executive, in particular in the context of the operationalisation of the Climate Action Plan. A rejuvenated PDA could be used as a way to catalyse these discussions. The PDA could, for example, identify key contact persons in the EPA and the departments for each area of work and how often they should meet. If it is judged that the PDA is not the correct tool, the EPA must explore other avenues to clarify expectations and improve the relationship Box 2.
- **Advocate** for more structured co-ordination and transparent communication with other regulators and public agencies. There could be value in defining regular structured engagement between management, for example: using Memoranda of Understanding to define how frequently the heads

of agencies should meet, and establishing MoUs with important partners where they do not already exist. Such co-ordination with other major actors could help identify common objectives and align regulatory activity.
- **Communicate** to the public and stakeholder groups which organisations are responsible for which areas of environment protection. For example, by improving the prominence, accessibility and readability of the "who does what" for Ireland's environment explanation, jointly with the other actors involved. This could be presented in corporate documents, on the website in an easily accessible location, and in important publications such as the State of the Environment report.

Box 2. Building constructive relations with the executive, the case of Great Britain's energy regulator Ofgem

The British energy regulator, Ofgem, has consciously tried to develop ongoing relationships with government without compromising its independence or giving the impression that there is undue influence. This has meant that where there has been a divergence of views, relationships have already been developed so that there can be dialogue. Ofgem has also tried to avoid surprises for ministers and officials – learning from adverse past experience.

One recent example is a review of network charging. This involved removing payments to some parties and rebalancing charges and making them fixed. The overall amount of money to be recovered is fixed so any rebalancing inevitably involves some parties paying more and some paying less. This created the risk that the government would come under lobbying pressure by those who would pay more, including in this case, larger businesses and renewable energy organisations. The major beneficiaries were average consumers, who by their nature are not as vocal as stakeholder associations.

Ofgem used its relationship with officials to ensure there was understanding of the basis for its reforms and to try to ensure that briefings for ministers reflected this and the consequences of not introducing the reform. There was also clear understanding of the independence of the regulator and the need to reflect this in any correspondence from ministers.

Source: Information provided by Ofgem, 2019.

Input

Financial resources and their management

More than three-quarters of EPA income is obtained directly from government sources, including from an Environment Fund that is designed to diminish but that is used to cover non-discretionary expenses, raising concerns about financial sustainability and autonomy. These funds require yearly support from DCCAE and are approved by the Department for Public Expenditure and Reform (DPER). EPA has two main sources of funding coming directly from government (Table 1):

- Exchequer income: Represented EUR 42.3 million (65%) of the EPA budget in 2018, with EUR 35.3 million from DCCAE and EUR 7 million from DHPLG. Funds from DCCAE support the overall functioning of the EPA, including staff, whereas DHPLG funds are dedicated to delivering priorities related to the Water Framework Directive. DHPLG funds can only be spent on staff when provided proper sanction to do so. Research is also partly funded by exchequer income.

- The Environment Fund: Collected by the national government from a plastic bag levy and landfill levy, and allocated by DCCAE. The amount of funding assigned from the Environment Fund has been declining from EUR 16 million (27% of the EPA budget) in 2015 to EUR 9 million (13% of the EPA budget) in 2018 while exchequer income has risen. The EPA does not have autonomy in the allocation of the Fund and in each of the last three years it has signalled in the Estimates Letters its concern about the sustainability of using this fund for non-pay, non-discretionary expenditures (i.e. operational costs related to light, heat, rent, insurance, etc.). This has been accompanied with requests each year to move these expenditures off the Environment Fund. In 2019, EUR 1.4 million of the Fund was allocated to such expenditures, down from EUR 4.6 million in 2018. However, the Fund also provides resources to carry out other important EPA functions, such as the remaining portion of research and waste prevention activities, and was an important source of funding during the austerity period.

Table 1. EPA budget by category, real (millions EUR) and percentage of total income

	2015	2016	2017	2018
Total income	59.0	59.8	63.0	65.0
Exchequer income	26.9 (45.7%)	33.5 (56.1%)	39.9 (63.4%)	42.3 (65.0%)
Environmental Fund income	16.1 (27.3%)	12.8 (21.4%)	9.8 (15.5%)	9.0 (13.8%)
Earned income	13.3 (22.5%)	10.7 (18.0%)	10.8 (17.2%)	10.9 (16.7%)
Other income	2.6 (4.5%)	2.7 (4.6%)	2.5 (3.9%)	2.9 (4.5%)
Total STATE Income	43.1 (73.0%)	46.3 (77.5%)	49.7 (78.8%)	51.3 (78.8%)
Total EARNED & OTHER Income	15.9 (27.0%)	13.5 (22.5%)	13.3 (21.2%)	13.8 (21.2%)

Source: Information provided by the EPA, 2019.

The remaining EPA budget is from earned income from levies from licencing fees, radiological income and enforcement income some of which are not set on a cost recovery basis putting pressure on other sources of funding, especially given the acceleration of economic activity in the country. Licensing fees are defined in legislation and require an amendment by the Oireachtas (Parliament) to change. These fees have rarely been reviewed since the EPA was founded. Generally, these fees cover 10-15% of the cost of licensing. In addition, there is no legal provision to charge fees for some licensing and permitting activities such as Article 27 notifications, EPA-initiated licence reviews (both of which have increased significantly since 2016) and EPA-initiated technical amendments. Enforcement income, however, is calculated on a near-recovery basis that includes the cost of the activity (including staff time and overhead costs) and chargeable costs associated with the activity. For those enforcement activities where costs are not fully recovered, the relevant team is asked each year to outline how they propose to charge for their activities for the coming year.

The EPA has a robust system of financial reporting, which has been well-recognised by external partners. Expenditures against the budget are reported to the EPA Board on a monthly basis, and internal budgets are reviewed bi-annually. DCCAE is provided monthly updates, and the EPA produces Annual Financial Statements audited by the Comptroller and Auditor General (C&AG) of Ireland, who reports to the Public Accounts Committee (PAC). The EPA can be called before the PAC to report on how it managed its resources. This has only occurred three times since its establishment. The most recent testimony in April 2019 received praise from the Chair of the PAC noting EPA's excellence in preparing its annual reports.

Recommendations

- **Secure** the sustainability of the financing framework. Developing stronger relationships and improving communication with parent departments, including with DPER, will assist with delivering more sustainable financing. The EPA should continue to advocate for a review of licensing fees set in legislation in order to better recover costs of activities and reflect the EPA's evolving role. The additional income could be used to fund improvements to the delivery of the licensing process. The increased financial autonomy could also strengthen the EPA's independence. In addition, the EPA needs to review the sustainability of financing coming from the Environment Fund, including planning ahead for alternative funding streams and continuing to advocate for allocating non-discretionary budgets to more stable budgetary streams.

- **Strengthen** co-ordination process with senior levels of DCCAE, DHPLG and DPER to raise, address and co-ordinate solutions to budget and human resource related issues.
- **Advocate** for stronger medium-term commitments to budgeting beyond the annual cycle, in line with the OECD Recommendation of the Council on Budgetary Governance. For example, the EPA could develop a fully-costed multi-year corporate plan, to be submitted and approved and then be subject to annual updates. This would allow the organisation to plan for expected budget availability.

Human resources and their management

As with all Irish public bodies, the EPA follows central government human resource frameworks and does not feel this impedes its ability to find and retain qualified candidates. The EPA is empowered to appoint staff subject to the numbers and grades sanctioned by DCCAE with the consent of DPER. Changes to the headcount require the approval of both bodies, which is usually only advocated for when new functions are added or there is growth in existing functions. Pay grades are similarly decided by central government rules. However, the EPA feels that it receives a large number of applications for each job posting allowing them to hire qualified candidates in most areas, with the exception being some posts in the Dublin Office due to higher costs of living, longer commute times and opportunities for other employment due to higher economic activity.

The EPA has developed an ambitious Human Resources Development (HRD) Strategic Framework and Action Plan 2017-2021 although it does not include measurable targets. The overarching theme for the HRD strategy is "Engaging, Enabling, Empowering" and it is supported by four strategic goals: foster a healthy, engaged, and resilient workforce; develop our people and organisational resources; empower our managers as experts and leaders; and evolve our HR delivery model. Each of the four goals is supported by four to five strategic priorities, as well as a high level outcome that the EPA intends to achieve by 2021, but these are not accompanied by measurable targets. Setting these targets provides an opportunity for the EPA to promote innovation and change management to ensure skills are fit-for-purpose as the role of the organisation evolves.

The EPA has workforce planning mechanisms that provide an opportunity for change as the organisation defines its future role. The EPA currently has 420 staff who are considered highly qualified in scientific and technical aspects by stakeholders. Around 70% of staff at the EPA are technical staff from engineering, science or specialised research backgrounds with the remaining considered administrative staff. The EPA does not currently have economists or lawyers on its staff. An annual workforce plan is prepared by HR in conjunction with directors; this plan also looks at options including outsourcing and contractors to deliver the required functions, which are highly concentrated in specialised areas such as ICT, chemicals, Pollutant Release and Transfer Register (PRTR) and legal services. As the EPA defines its future identity and role, the workforce plan will be a powerful mechanism to translate the new vision into the capacity to deliver on it.

The EPA uses a system of lateral mobility to develop skills and maintain internal flexibility, which has been seen as creating positive and – in some cases – negative results. In general, EPA staff are assigned to a position upon appointment but may be transferred to new assignments for organisational and/or development purposes. The EPA Performance Management and Development System (PMDS) is used in some cases to match positions to candidates for lateral mobility opportunities. Lateral mobility is either initiated on a voluntary basis by the staff member or by management to broaden staff skills and redeploy resources where needed. This system has been seen by staff as beneficial to development. In cases where moves are management-initiated, they may be perceived as top-down.

Recommendations

- **Align** skills with the mandate and strategic direction of the EPA. Whatever strategic direction the EPA decides to pursue, the skills of the organisation must follow suit. The EPA could conduct a foresight exercise on the types of skills and backgrounds likely to be needed and then make the business case for the creation of new specialties and teams to help position the organisation for future growth: for example, hiring appropriately qualified in-house legal resources given the context of an increasingly litigious environment; or policy professionals to advise the Board if it moves more into the policy space, etc.
- **Engage** regularly with DCCAE, DHPLG and DPER to ensure that the EPA has sufficient human resources in line with the assessment of the skills and posts needed for the organisation to deliver on its mandate.
- **Harness** the PMDS to ensure that the EPA's strategic objectives are reflected at all levels of the organisation. As well as identifying training needs, staff appraisals should be used to evaluate potential candidates to be the next generation of leaders of the organisation. Complementary systems could be used to provide opportunity to practise leadership skills and receive mentorship, such as the Management and Leadership Network (MLN) established under the new HRD strategy.
- **Ensure** the attractiveness of EPA as a modern employer. The HRD Strategy could be a potentially powerful tool in modernising HR practices within the EPA but it will be important that measurable targets are attached to the goals (Box 3). The EPA could also invest more time in communicating and gaining support from staff identified as candidates for management-initiated lateral moves. The EPA could consider new approaches in this area such as advertising such positions internally.

Box 3. Performance indicators for people management from the Treasury Board of Canada Secretariat

The Treasury Board of Canada Secretariat (TBS) introduced a people management methodology in 2017-18 in order to provide a portrait of the health of an organisation in terms of its management practices and performance with respect to people, structures, processes and well-being.

The key areas of assessment were:

1. *Workforce:* includes measures related to talent and performance management, learning and development, and official languages.
2. *Structures and processes:* provides a picture of how each organisation is structured (in terms of executive population, levels of executive reporting to direct managers, etc.). These measures also give an indication of how the organisation is designed and if the organisation works effectively to meet changing job demands while ensuring that jobs clearly reflect the work to be performed.
3. *Workplace culture:* includes measures on mental health and wellness, diversity and inclusion, and values and ethics.

The methodology defined indicators and their calculation methods matched with expected results (targets) for each area of assessment (Table 2). To the extent possible, and with a view to leveraging existing information, the indicators were aligned with the goal of a healthy and productive workforce.

Findings from existing employee surveys were incorporated into the analysis of results from central system data and requests from departments.

Table 2. Extract of the questionnaire on Workforce

Workforce

Outcome statement: A public service that enables new and existing public servants to be in the right place, at the right time, doing the right things.

Talent and performance management, learning & development

Outcome statement: A skilled and agile workforce that has the competencies and flexibility to meet the needs of an evolving public service.

Rationale: A world class public service equipped to meet the challenges of the 21st century requires continuous learning, training and professional development for employees and executives. An organisation's commitment to various ways of learning is the foundation of employee development and performance improvement.

Indicators and calculation method (where applicable)	Expected result
1. Percentage of employees that have documentation setting performance expectations/objectives. **Rationale:** To measure the extent to which departments and agencies meet the requirements of the TBS directive on Performance Management. **Calculation:** Number of employees who have documentation setting performance expectations (objectives) ÷ total number of employees × 100%. **Employee tenure:** Indeterminate and term employees of more than 3 months (non-Executives). **Employee status:** Active employees	Organisations should strive to have over 90% of employees with documentation setting performance objectives
2. Percentage of executives that have documentation setting performance expectations/objectives. **Rationale:** To measure the extent to which departments and agencies meet the requirements of the Directive on Performance Management Program for Executives. **Calculation:** Number of executives who have documentation setting performance expectations (objectives) ÷ total number of executives × 100%. Employee tenure: Indeterminate and term executives of more than 3 months. Employee status: Active executive employees	Organisations should strive to have 100% of executives that have documentation setting performance expectations/objectives.
3. Percentage of employees that have documentation setting learning objectives (learning and development plan). **Rationale:** To measure the extent to which departments and agencies meet the requirements of the TBS directive on Performance Management. **Calculation:** Number of employees who have documentation setting learning objectives (learning plan and development plan) ÷ total number of employees × 100% **Employee tenure:** Indeterminate and term employees of more than 3 months (non-executives) **Employee status:** Active employees	Organisations should strive to have over 90% of employees with documentation setting learning objectives.

> For the full list of indicators, see
> https://www.canada.ca/en/treasury-board-secretariat/services/management-accountability-framework/maf-methodologies/maf-2017-2018-people-management-methodology.html#question.
>
> The results were intended to provide information to three key audiences:
>
> - Deputy Heads, so that they could identify the strengths and potential risks in their organisations in relation to corporate commitments, such as talent management, diversity and inclusion, and well-being; and provide information to track and communicate progress on the Government of Canada's people management priorities.
> - People Management Community, to measure the effectiveness of human resources services and identify areas of strong performance as well as gaps.
> - The Treasury Board Secretariat, to enable policy centres to monitor trends and identify gaps across departments and enterprise-wide relating to government priorities; and support TBS program sectors and departments with evidence-based analysis on departmental initiatives.
>
> Source: https://www.canada.ca/en/treasury-board-secretariat/services/management-accountability-framework/maf-methodologies/maf-2017-2018-people-management-methodology.html#question.

Process

Decision making and governance structure

The EPA is managed by a full-time Executive Board that must fulfil executive, management and strategic duties. The Board comprises the Director General and five directors, who are appointed by Government after an open competition. The Director General serves as Chair of the Board and operational chief executive of the EPA. Each director leads an Office (internal department of the EPA) and provides day-to-day oversight of the EPA. Legislatively, the Board has responsibility for the management of the EPA but it is empowered to delegate responsibility to staff for operational purposes. Currently twelve programme managers are delegated operational responsibility for carrying out the work of the EPA. The governance arrangements set out in the EPA's founding legislation take precedent over the Irish Government's *Code of Practice for the Governance of State Bodies 2016* (DPER, 2016[1]) which states that the role of Chairperson and CEO should not normally be combined, and that non-executive Board members should bring an independent judgement to bear on issues of strategy, performance, resources, key appointments and standards of conduct.

EPA leadership has implemented a number of measures in recent years to strengthen and diversify strategic decision-making; continuing these efforts is key in the current changing policy context. Most meetings of the Board are technical in nature (i.e. concerning complex licensing decisions, prosecutions), with one meeting per month dedicated to monitoring the delivery of the work programme, corporate governance and strategic matters. While the Board functions smoothly with regard to its executive functions, the current weekly schedule does not allow sufficient time and focus on strategic matters. Therefore, in addition to regular Board meetings, Board members and senior managers (a group of 18 people) meet at least six times a year to discuss the EPA's strategy and its implementation, although decision making remain at the Board level. To support this process, a series of external experts have been contracted to support the Director General on topics such as change management and organisational development. Guaranteeing sufficient resources for the steering of the EPA's strategic vision as well as strategic decision making will be essential for the organisation to navigate its way through a changing policy context in Ireland where environmental issues and in particular climate change are rising up the political and public agendas.

The current governance arrangements and wide responsibilities of Board members highlight the importance of seeking diverse external perspectives to strengthen decision-making. The EPA Act sets out that the role of the Chair and CEO is occupied by one person and all Board members are part of the management team, which can lead to a lack of distance from EPA operational and technical matters. The lack of non-executive directors on the Board also limits the level of external input and challenge. Moreover, Board members have generally worked in the EPA for a number of years (currently an average of 16 years), which may run the risk of 'group think' and a lack of new perspectives. To mitigate against these risks, the EPA engages external expertise to support the Board functions, including financial, procurement, governance, legal, HR and communications; external experts can attend Board meetings. Various internal and external committees currently advise the Board and the scope for them to do so may be currently under-utilised. A current strength that could be further utilised is the Audit and Risk Committee (ARC) that consists of primarily external members and is externally chaired. The role of the ARC is to provide independent assurance to the Board on the effectiveness of the control environment, risk management and the internal audit function. The Chair of the ARC attends the Board of the EPA at least once per year and prepares an annual independent report which is presented to the Board.

The Advisory Committee also holds potential that could be further exploited. The Advisory Committee is empowered in legislation to make recommendations related to the functions of the EPA to the EPA and to the Minister, yet does not do so frequently. Standard practice has been for the Advisory Committee to draft a single report with recommendations at the end of its term. The role of the EPA Director General as Chair of the Advisory Committee, including setting meeting agendas in consultation with members also reduces its potential to leverage external perspectives. Large amounts of meeting time is given to presentations from EPA staff on various topics, rather than being used for discussion and comments.

Recommendations

- **Continue to strengthen** the strategic function of the Board. This could be done in a number of ways, such as:
    - isolating opportunities to discuss strategic matters, for example, holding a quarterly Board retreat dedicated to strategy (Box 4);
    - inviting external input at Board meetings to help bring innovative, 'blue sky' thinking to aid the Board in its strategic discussions;
    - consider creating a "strategic advisor" post to support the Director General/ Board.

Box 4. Business Planning and Strategic Oversight at Canadian Transportation Agency

Like many organisations, the Canadian Transportation Agency (CTA) has a senior-level group for the discussion of, and decisions regarding, management and administrative matters: its Executive Committee (EC), which includes the CTA's Chair and CEO, Vice Chair, and the heads of all branches. EC meets weekly, but the primary forum for the establishment of organisational plans and priorities is its quarterly, day-long retreats.

These retreats provide an opportunity for the senior executive team to step back from operational pressures, creating space for reflection and open conversations on:

> - *Long-term trends, opportunities, and challenges:* identifying key trends and best practices in the external environment and significant issues in the organisation – based in part on thought-provoking presentations from expert staff and invited guests – and considering their implications for the delivery of the Agency's mandates;
> - *Major project updates:* sharing information on major initiatives, to ensure transparency, accountability, the collective monitoring of progress, and early action in response to any challenges;
> - *Organisational priorities and resource allocation:* determining areas of focus and investment, and re-aligning budgets as required to achieve results and mitigate workload pressures;
> - *Talent management:* discussing options for addressing performance gaps and for developing and leveraging the competencies of exceptionally strong employees.
>
> These retreats allow senior officials to think strategically about organisational directions and to respond to shifting demands and emerging issues – resulting in a strong sense of common purpose and highly effective, agile decision-making. Their outcomes are communicated to staff at branch-level meetings and through the Hub, the CTA's intranet site, and are implemented and tracked through planning documents, people management processes, and follow-up discussions at the EC table.
>
> Source: Information provided by CTA, 2019.

- **Diversify** input into decision-making by continuing to find opportunities to bring in external perspectives, fresh ideas and constructive challenge. Strengthening the links between the Board and its external committees could be one way to achieve this.
- **Clarify** the role of the Advisory Committee and redesign meeting formats to produce greater value.
    - Continue the practice of co-creating meeting agendas, looking for opportunities where the Advisory Committee's input could strengthen EPA performance, for example: giving comments on draft standards, guidelines and codes of practice, strategic plans, the research programme, and on potential EPA responses to wider policy developments (e.g. the release of government white papers, environment-related plans etc.) and environmental challenges.
    - Redesigning the format of meetings to give more time for discussion and comments could provide much more value. For example, relevant documentation (presentations/reports/podcasts etc.) could be sent for members to consult in advance of each meeting, freeing up valuable time.
    - The Committee's mandate to make recommendations to the EPA and to the Minister could be better fulfilled by changing the format of reporting. Rather than publishing a single report of recommendations at the end of its three-year term, the Advisory Committee could consider whether there would be value in issuing recommendations to the EPA Board and to the Minister following discussions on specific topics or issues as relevant. This potential could also serve to focus discussions.
- **Continue** efforts to increase transparency and visibility of Board and sub-committee meetings. For example, publish minutes on the website in an easily accessible location (excluding any information that could be commercially sensitive or confidential internal issues, e.g. human resources) (Box 5).

Box 5. Transparency at Mexico's National Hydrocarbons Commissions

Mexico's National Hydrocarbons Commission (Comisión Nacional de Hidrocarburos, CNH) implemented several initiatives to improve transparency inside the regulator.

First, CNH has been transparent in making information on its decision-making processes available, including publishing the minutes, resolutions and technical support documents of the governing council meetings on the CNH website, and the meetings are streamed live and archived on the Internet.

Second, in October 2016, the Natural Resource Governance Institute (NRGI) presented the report on "International Best Practices for Transparency in Contract Management: Recommendations to the National Hydrocarbons Commission of the Government of the United Mexican States" that reviews and evaluates the transparency of the CNH website.

Based on methodology presented in the report, the CNH created a transparency group made up of NGO representatives and academic institutions (full list can be found in (OECD, 2017[2]). The objective was to enhance participation and facilitate stakeholder monitoring that would allow incorporating and addressing their own concerns, information needs and recommendations to strengthen transparency in bidding processes, and contract management. The interaction with this group has allowed CNH to identify gaps in transparency matters and to carry out actions to improve processes and make relevant, clear and accessible information available to society.

As a result, CNH now ensures that information on the process of each bidding round is public and online on its revamped website—contracts, annexes, the name of companies involved at each stage of the bidding process. Progress in the bids can now be tracked by external parties. Contracts with foreign oil companies have also been made fully public, as have companies' investment plans.

In June 2018, the NRGI and the Open Contracting Partnership (OCP) published the report "Open Contracting for Oil, Gas and Mineral Rights: Shining a Light on Good Practice". The report studied 14 countries in which Mexico stands out in seven of the 16 best international practices of transparency in bidding processes and contract administration, thereby positioning CNH as a leading institution in this area.

CNH has strengthened its engagement and transparency commitment with regard to civil society and academia, by creating the Monitoring and Transparency Group and agreeing to comply to jointly agreed information requirements on Exploration and Extraction contracts, entitlements and compliance with exploration plans. CNH also participates in NRGI's initiative to evaluate Entitlements, Contracts and Implementation of the bidding process in terms of the Extractive Industries Transparency Initiative (EITI) standard or to develop Open Government Partnership (OGP) national action plans and in November 2018, CNH accepted EITI's invitation to lead the international network for contract transparency.

Furthermore, some of the functions of CNH are to collect, preserve, manage, analyse and update the information belonging to the Nation, obtained from survey and surface exploration activities, as well as hydrocarbons exploration and extraction activities, which are carried out through the National Hydrocarbons Information Center (CNIH). This centre manages an Internet platform with statistics, maps, data and reports for the public and is also subject to suggestions for improvement by the Monitoring and Transparency Group.

Source: (OECD, 2017[2]), Driving Performance at Mexico's National Hydrocarbons Commission, The Governance of Regulators, Paris, http://dx.doi.org/10.1787/9789264280748-en; OECD (2018), "Impact Update: Driving Performance of Mexico's Energy Regulators", OECD, Paris. http://www.oecd.org/gov/regulatory-policy/ner.htm and information provided by CNH, December 2019.

Internal organisation and management

Functions and subject areas are currently split between EPA Offices and locations, creating challenges in terms of efficiency, consistency of approach and messaging (Figure 1 and Table 3). Licensing functions are shared between the Office of Environmental Sustainability (OES) and the Office of Radiation Protection and Environmental Monitoring (ORM), which carries out licensing and enforcement functions for activities relating to radiation. Enforcement is shared between the Office of Environmental Enforcement (OEE), OES and ORM. This separation of functions between offices may lead to a disjointed approach. Work on certain themes is also dispersed across the EPA, for example, teams working on air quality are located in ORM and OEE, and climate change teams are in OES and the Office of Environmental Assessment (OEA). In many ways, the fragmentation that is characteristic of Ireland's environmental legislation is reflected in the EPA's own internal organisation. The responsibilities that have been given to the EPA over time have been assigned to different teams, leading to a mosaic of functions within each Office.

Figure 1. EPA organisational structure

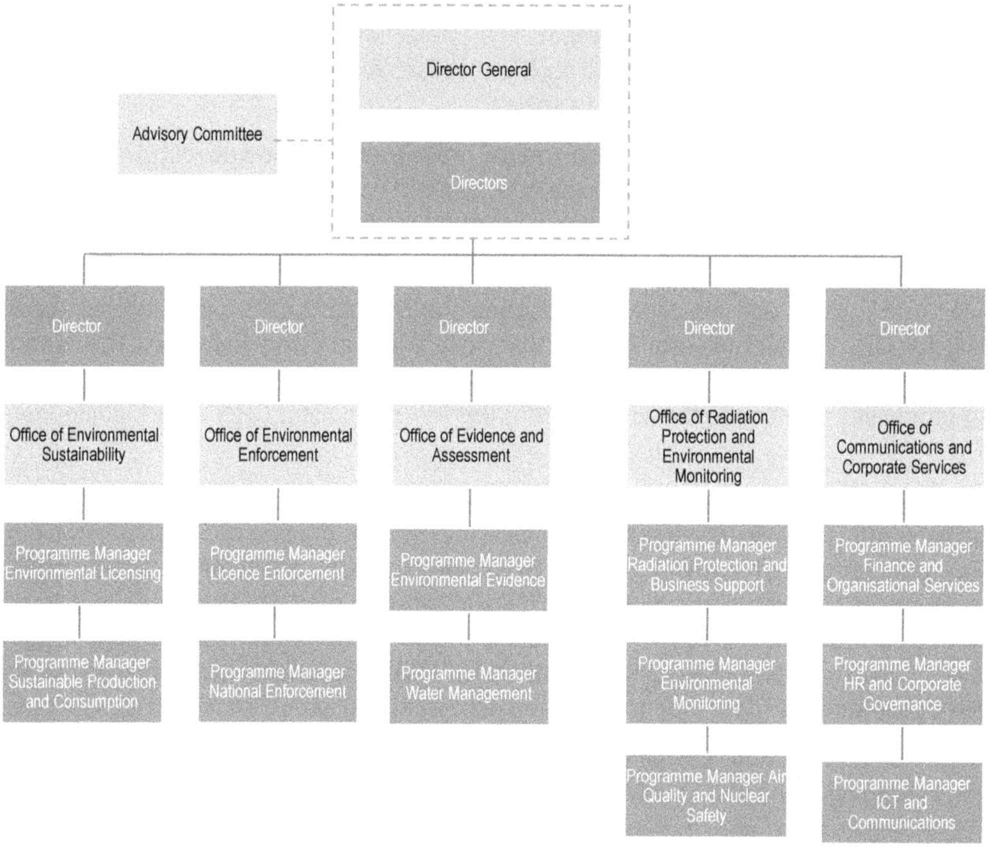

Source: EPA website https://www.epa.ie/about/org/.

The EPA is a decentralised agency which supports proximity with regulated activities but may at times hinder efficiencies and effective internal and external communications. While the majority of staff are based in two main locations – the headquarters in Wexford (150 staff) and a regional inspectorate in Dublin (120) – the EPA is present in six other locations: four regional inspectorates in Castlebar (30), Cork (50), Kilkenny (20) and Monaghan (14) and two smaller offices in Athlone (2) and Limerick (2), with water testing laboratories in four of these locations. Regional inspectorates include staff from across several Offices. There appears to be a high degree of travel involved day-to-day, especially for senior management, which hinders efficiency and increases the carbon footprint and environmental impact of the organisation, sitting at odds with the overall goals of the EPA. For specific functions, however, the regional presence can be an advantage for the EPA: for example, being close to licensees facilitates enforcement activities.

Structures put in place to improve internal communication and overcome the challenges posed by the regional structure show promise, but silos remain. A Senior Management Network (SMN) that brings together directors and programme managers across all offices is a significant development in the last three years that appears to be working well and is positively regarded by staff. An expert in culture and team effectiveness is engaged on an ongoing basis to work with the Director General, the Board and the SMN on team dynamics and performance, culture and corporate performance. Other initiatives include cross-office groups on specific topics and meetings of technical functions. Despite this, examples of good practice and innovative approaches developed by individual Offices seem to remain within their silos rather than being mainstreamed across the organisation; for example, on waste activities or policy submissions.

Table 3. EPA's five offices and their functions

	Functions	Themes
Office of Environmental Enforcement (OEE)	Enforcement Compliance promotion Inspections Monitoring Advisory (to local authorities) Produce guidelines	Air quality Waste Drinking water Wastewater VOCs Financial provision
Office of Environmental Sustainability (OES)	Licensing and permitting Enforcement Environmental impact assessments Compiling national data and reporting Advocacy and public information Applying behavioural insights National plans	Sustainable production and consumption, the circular economy and waste Climate change Air quality Waste Waste water VOCs GMOs Greenhouse gas emissions Peat extraction Chemicals
Office of Evidence and Assessment (OEA)	Strategic environmental assessments Monitoring Reporting (e.g. state of environment report) Data analytics (service to other offices) Advocacy and public information (public lectures series, specialised websites etc.) Co-ordinates research Climate services (e.g. Secretariat to CCAC) Water management, river basin management (WFD)	Water quality Hydrometrics Climate change
Office of Radiation Protection and Environmental Monitoring (ORM)	Radiological licensing and enforcement Product certification Monitoring (air quality, radiation levels in water, soil and food), modelling and forecasting Laboratory services Advice and public information (radon, non-ionising radiation) Advocacy (e.g. citizen science) Emergency preparedness plans Radiation research	Radiation Nuclear safety Environmental emergencies Air quality Radon Non-ionising radiation + "Clean air, clean water, sustainability" (themes of the citizen science initiative)

	Functions	Themes
Office of Communications and Corporate Services (OCCS)	Corporate functions (Human resources Corporate governance ICT and communications Finance Organisational services)	Cross-cutting

A heightened focus on internal control and risk management has been placed on all State bodies, and the EPA is in the process of improving its system for evaluating risks. In accordance with *Code of Practice for the Governance of State* Bodies, the EPA Board is responsible for ensuring that effective systems of internal control are instituted and implemented. Updates to the Code in 2016 placed a heightened focus on risk management and requires all State bodies to have an Audit and Risk Committee (previously the Audit Committee). In 2017, the EPA commissioned an independent external gap analysis of EPA compliance against the Code. The evaluation identified that "the Agency has already made commendable efforts in complying with the provisions in the revised 2016 Code". The EPA audits its compliance and control four to five times per year, and its compliance record is considered very high. The Executive Risk Committee (ERC) is responsible for further internal control through a Corporate Risk Register that is currently being updated to identify a more focused list of risks (12 as opposed to 33) and consider the likelihood and impact of risks, which the previous version did not. While this is a considerable improvement, the register could further benefit from being directly linked to the EPA's organisational strategy.

The EPA invests significant time and resources in carrying out internal reviews, but the extent to which the output from reviews contributes to organisational improvements is less clear. The EPA has undertaken 17 reviews in the last 10 years, some of which have been in response to requirements in the Code of Practice 2016. It is not clear whether the EPA uses the results of reviews to continually improve performance. For example, actions from previous reviews have not all been accepted and delivered.

Recommendations

- **Streamline** the internal structure for efficiency and cohesiveness gains. The EPA could benefit from taking stock of the distribution of functions across Offices with the goal of bringing together functions that are currently dispersed. In particular, there is a strong case for centralising enforcement functions into one Office. There also seems to be scope and appetite to bring together work on thematic areas that are currently divided between Offices, notably the work on climate change. A process redesign could be the first step on the way to a new organigram, to ensure uniformity and standardisation of processes.
- **Continue to review and assess** the efficiency of the EPA's regional structures. The EPA's multi-site operation warrants ongoing review and, where possible, roles should be brought together as closely as possible. While regional presence can be a strength for certain functions (e.g. enforcement), in other cases the current set up could be assessed. For example, an update of the previous reviews of the business case for the regional laboratory structure may be worth considering (Box 6).
- **Link** the corporate risk register more directly to organisational strategy. The more focused list of corporate risks, using an approach that considers the likelihood and impact of risk, is a significant improvement on the previous version.
- **Use** reviews as learning opportunities to continually improve performance. This will require continuing efforts to foster an organisational culture that is more open to change. Previous reviews could be revisited and assessed independently to see whether they still have relevance and how they should be taken forward. Priority can be given to reviews according to risk.

> **Box 6. Water quality testing in Portugal**
>
> Portugal's Water and Waste Services Regulatory Authority (ERSAR) approves and supervises the water quality testing programmes (PCQA) for all water suppliers, as required by legislation (Portuguese legislation based on the transposition of the Drinking Water Directive 98/83/EC).
>
> As stipulated in the PCQA, water suppliers are responsible for monitoring water quality in their supply zones and must choose a laboratory that has been approved by ERSAR to carry out the analyses.
>
> By law, ERSAR is the supervisory body for laboratories responsible for the quality testing of water intended for human consumption. For that purpose, ERSAR assesses the credentials of the laboratories and publishes a list of accredited laboratories from which water suppliers must choose. ERSAR does not operate any laboratories itself.
>
> Source: Information provided by ERSAR.

Regulatory processes

The EPA demonstrates independence in its regulatory functions. Licensing and enforcement decisions appear well insulated from external interference and the EPA takes pride in its technical independence, bolstered by provisions within the EPA Act that make it an offence to try to influence improperly any employee of the EPA or its advisory committees. EPA leadership also highly value and guard the independence of their regulatory decision-making.

Licensing and permitting

Licensing and permitting are conducted through a transparent but detailed process, which is considered by the European Environmental Bureau as ranking (along with Norway) the best in Europe (EEB, 2017[3])**. There may be room to further streamline the process without impacting rigour.** Licences and permits are granted by the EPA across a range of sectors (Table 4). The process involves a transparent online portal where applications are received, submissions and related information are made available to the public. Both the public and the prospective licence holder can also submit comments at various stages of the process. The licence and permitting process takes on average 1.5 years to complete, while the EPA seeks to bring this down to nine months.

Table 4. Responsibilities for licensing/permitting and enforcement at the EPA

Category	Licensing/permitting responsibility	Enforcement responsibility
Waste facilities	OES	OEE
Large-scale industrial activities	OES	OEE
CO_2 emissions trading	OES	OES
Intensive agriculture	OES	OEE
Air quality	Registration with EPA	Local authorities
Genetically Modified Organisms (GMOs)	OES	OES
Drinking water by public water suppliers	n/a	OEE
Waste water discharges	OES	OEE
Dumping at sea	OES	OEE
Sources of ionising radiation	ORM	ORM
Large petrol storage facilities	OES	OEE

Category	Licensing/permitting responsibility	Enforcement responsibility
Local Authorities	n/a	OEE
Waste Electrical and Electronic Equipment (WEEE)	OES	OES
Chemicals	OES	OES
Volatile Organic Compounds (VOC) Permits	OES	OEE

Notes: OES = Office of Environmental Sustainability; OEE = Office of Environmental Enforcement; ORM = Office of Radiation Protection and Environmental Monitoring.

There appears to be little flexibility to increase resources for certain licensing when the workload increases as a result of new categories of licence being introduced, when updates to existing licences are required, or as the economy expands. For example, the EPA makes revisions to all affected licences when new Best Available Techniques (BATs) for licences are confirmed. BATs seek to use the most effective and advanced activity and method of operation to achieve a high level of protection for the environment. While the requirement to review and revise licences when new BATs are confirmed is an excellent practice, it can cause a significant pressure and delay in the licensing process if adequate resources are not allocated. Looking forward, a new Water Framework Directive requirement is anticipated for licensing large water abstractions. Estimates forecast that this will require 800 new licences, representing a substantial volume of work for the EPA. To meet this demand, the EPA requested and received sanction for nine additional posts that cover hydrometric, water abstraction licences and reviews of wastewater licences. Requests for additional posts for other licensing areas have not been sanctioned. Stakeholders strongly advocate for more pathways and faster revision processes, especially for new directives.

Inspections and enforcement

The EPA is responsible for inspections and enforcement activities for the licences and permits that it grants but in some sectors, such as waste, there is fragmentation of enforcement responsibilities in the state. Overall, the EPA is responsible for inspecting and enforcing approximately 830 industrial and waste licences. In the waste sector, the EPA is responsible for large-scale waste facilities. Permitting, inspections and enforcement for smaller waste facilities are under the responsibility of local authorities. The EPA has a dual role with regards to local authorities. On the one hand, they are responsible for supervising the enforcement actions of local authorities. On the other, the EPA is responsible through the NIECE network for supporting the co-ordination of a consistent and effective approach to the enforcement of environmental legislation by local authorities and others in Ireland.

The overall inspections and enforcement strategy is compliance-focused and takes into consideration risk, but more can be done to improve this approach by further focus on outcomes. The EPA is finalising a Compliance and Enforcement Policy that establishes high level policy goals, which has been approved by the Board and set for implementation in 2020. This policy is in line with the OECD Best Practice Principles on Enforcement and Inspections. The EPA collects, assesses and reports on the number of non-compliances, new compliance investigations opened, incidents, number of site visits conducted, number and amount of prosecutions, and complaints received from citizens. These are then used to inform and prioritise enforcement action. In some cases, priority actions lists are used to promote compliance, such as the Remedial Action List (RAL) that seeks to increase investment to improve public drinking water. Using numerical rather than outcome measures to decide and report on inspections and enforcement is generally not in alignment with principles of responsive regulation and can provide perverse incentives that lead to more inspections and enforcement actions than is often optimal and for offences that are proportionally less risky. By focusing on outcomes, through a risk-based approach, the EPA can better utilise its range of enforcement tools to encourage compliance (Figure 2).

Figure 2. Types of EPA enforcement actions

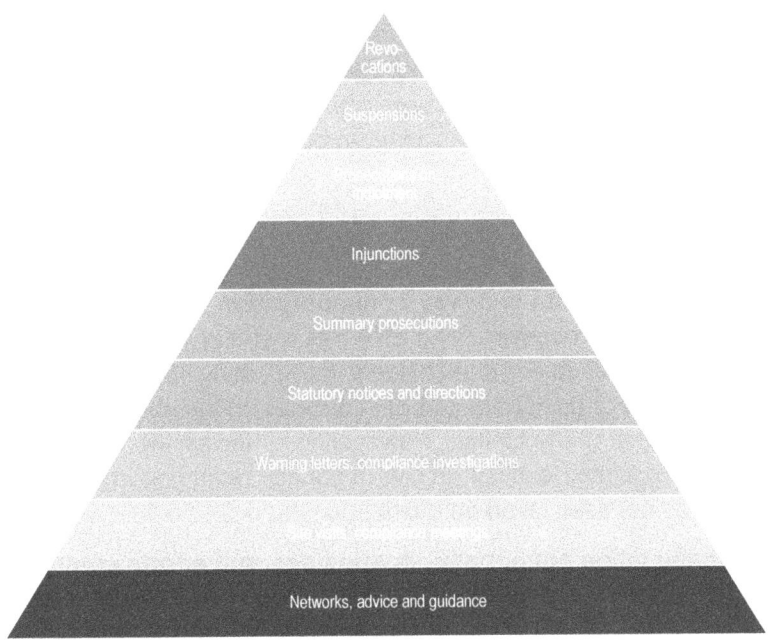

Source: 2017 EPA Industrial and Waste Licensing Enforcement report
https://www.epa.ie/pubs/reports/enforcement/EPA_Industrial_Waste_LE_Report2017.pdf.

The EPA possesses a variety of sanctioning powers intended to achieve compliance, which, if used consistently in line with a risk-based approach, should ensure efficient enforcement. The EPA has a variety of compliance and enforcement tools, ranging from advice and guidance to suspension or revocations of licences, which are in line with OECD Best Practice Principles for Enforcement and Inspections. The EPA can use prosecution, but fines are capped and may be too low to induce real behaviour change. The EPA should ensure that its existing range of sanctioning powers are employed in alignment with a risk-based approach to avoid the risk of launching compliance investigations on infractions that are of low risk to the environment.

The National Priority Sites (NPS) system is an innovative system to promote compliance through behavioural change and could be further strengthened. The National Priority Sites is a list of the worst environmental offenders, which attempts to "name and shame" sites into compliance. The list is updated quarterly based on data collected over the previous six months and has been initially successful: a review of companies on the list in 2017 showed that 14 companies stayed on the list one out of four quarters and only one site was listed for all four quarters. Rankings are decided according to a combination of four enforcement factors: complaints, incidents, compliance investigations and non-compliances with the licence. While some of these factors are based on risk or outcome measures, other factors could be susceptible to biases in perception, such as complaints that can account for up to 20 of the 31 points necessary to be on the NPS list when tied to a medium or high compliance investigation. This opens the possibility for perverse incentives for some to "game the system" by artificially inflating scores with potentially erroneous complaints that could result in a large impact on a regulated entity's overall score. The NPS system was reviewed in 2019, with amendments to the ranking formula approved by the Board in September 2019 for implementation in 2020. This includes reducing the weight of complaints in the overall score from 30 to 20. While not removing the risk of gaming entirely, it helps to minimise some of its effects. Continuing to review the system with a focus on ensuring robust verification procedures are in place can help safeguard this innovative system.

The EPA engages in a number of networks aimed at sharing information but operational co-ordination on enforcement and inspection activities is less systematic. The EPA participates in networks aimed at sharing information about enforcement, such as the Network for Ireland's Environmental Compliance and Enforcement (NIECE), the National Waste Enforcement Steering Committee and Waste Enforcement Regional Lead Authorities. It also liaises with a number of state bodies with responsibility for inspections that overlap with the EPA. On occasion, co-ordinated site visits are planned. However, this is not systematically organised.

Stakeholder engagement

The EPA should seek to improve its two-way communication with citizens and stakeholders to enhance trust in its regulatory processes. The EPA has a strong internal commitment to adhering to the Aarhus Convention, including engaging stakeholders in evaluating licence applications and developing new guidelines and processes. For guidelines, the EPA decides on the nature of the consultation process depending on the significance of the guidance. However, it is unclear what defines this threshold. It is important that citizens and regulated entities receive feedback regarding their comments, especially in cases where a comment has not been taken on board.

Avenues for general consultation are not always apparent, and efforts could be made to promote more systematic early stage consultation. The EPA draws on several avenues for stakeholder consultation. The EPA engages with a number of networks, which include state, local business and NGO actors. One of which is the Irish Environment Network (IEN) composed of environmental NGOs, which the EPA meets with on a biannual basis. Six external committees exist, including the Advisory Committee that meets most often. While these committees are periodically used to receive advice, they could be used more systematically when developing new guidelines and codes of practice. Furthermore, the consultation portal for general consultations on the EPA website is located in a difficult-to-find location, which may result in fewer responses.

Recommendations

- **Seek** and implement opportunities to streamline the licensing process to drive down the processing time to the organisational goal of nine months. This can include:
    - Explore the scope to standardise licensing to help address resource issues in this area, for example, making greater use of standardised templates. The EPA could also evaluate whether elements of licences can become 'approved by default' pending a full evaluation by a future inspection to further speed up approvals processes.
    - Invest in making licensing as transparent as possible by making guidance on as many commonly required elements for licences publicly available, and rewarding adherence with this guidance to speed up processing times. Consider creating a database of currently-approved practices to give entities clarity on how previous cases were resolved and guide solutions for future projects, as well as drive consistency among inspectors.
    - Communicate strongly with stakeholders on where to find streamlined information on licensing – including posting in a central location on the EPA website – as well as making available members of the inspections team to receive advice, in an effort to help licensees be initially compliant and hence speed up the approvals processes.
- **Document** precisely and communicate clearly to the executive the resources that are required for licensing activities in periods when the workload increases significantly, once any process efficiency gains have been exhausted. In case additional resources are not available, engage with parent departments to discuss and agree in a transparent manner the EPA's priorities, in order to ensure that resources are allocated accordingly.

- **Advocate** for a new licence review process based on the principles of proportionality and risk, including creating one or more avenues between a full review and technical amendment for revisions reflecting varying degrees of risk to the environment. This should especially be utilised when new standards come into force, such as new BATs, that require a large number of revisions.
- **Fully implement and monitor** with performance indicators the new Compliance and Enforcement Policy to ensure that the enforcement and inspections system adheres to the principles of responsive regulation, as elaborated in the OECD Best Practice Principles for Enforcement and Inspection. These can include the following actions that can help promote compliance with regulations:
    - Ensure enforcement is consistently applied according to a methodology based on risk and proportionality that takes into consideration the probability and consequence to the environment, alleviating the need to sanction smaller scale infractions and promoting good practice. The approach of other EPAs may provide inspiration in this regard (Box 7) which could be augmented with research conducted through the EPA's research programme or work applying behavioural insights.
    - Develop detailed guidelines or checklists for regulated entities to follow in order to be as compliant as possible.
    - Ensure inspectors are trained with all skills necessary to deliver on compliance-focused inspections, including substantial training not only on technical but also other necessary professional skills such as understanding and managing risks, communicating and advising, promoting and supporting compliance, investigation, etc., as recommended in the OECD Regulatory Enforcement and Inspections Toolkit;
    - Work with regulated entities to identify the most burdensome inspection requirements.
- **Review** the scoring system behind the National Priority Sites to ensure all points can be verified before determining whether a site should be placed on the list to ensure fair treatment. For instance, public complaints could be investigated prior to adding them to the tally or complaints can be aggregated by issue (i.e. one complaint is registered for one issue, no matter how many complaints are received) to avoid gaming behaviour. Similarly, ensuring the right categorisation of compliance investigations is confirmed via internal review prior to adding to the tally can help minimise any potential individual errors.
- **Systematically co-ordinate** where possible and practicable with other state agencies who inspect the same licensees to try as much as possible to co-ordinate visits, information collected and compliance requirements, as recommended in the OECD Regulatory Enforcement and Inspections Toolkit. Extend this co-ordination to state agencies not involved in the environment to exchange information about regulated entities, which may reveal more systemic violations. Seek updates, where necessary, to legal frameworks to enable such activities.
- **Work** with local authorities and DCCAE to clarify the EPA's role as both providing advice to and supervising compliance of local authorities to ensure the relationship remains fit-for-purpose over the long-term. Continue the good work of the NIECE network for co-ordination, strengthening the focus on compliance assurance by local authorities. For example, it could be used to design processes and identify common rules and procedures for undertaking inspections.
- **Engage** in two-way conversations with stakeholders to improve trust in regulatory processes. Proactively and constructively engage with regulated entities when developing guidelines, reviewing internal processes (e.g. licensing processes) etc. Align to best practice in consultation by placing the consultation portal in a clearly visible location and systematically providing feedback to comments received: after each public consultation, provide a report (publicly disclosed on the EPA website) with comments about all the feedback provided by entities/citizens. The report should explain which recommendations were accepted by EPA and which ones were not accepted and

why. Such engagement need not jeopardise the EPA's independence from industry if appropriate structures are in place that foster a culture of independence.
- **Investigate** ways to improve the structures and mandates of the other external committees to ensure that the EPA is receiving holistic and systematic early stage advice from stakeholder groups.

Box 7. Principles of responsive regulation applied to promoting compliance

The OECD Regulatory Enforcement and Inspections Toolkit recommends that enforcement should be based on the principles of "responsive regulation," meaning that enforcement actions should be modulated depending on the profile and behaviour of specific businesses. This means focusing on the core goal of achieving compliance with regulations by foreseeing a range of differentiated responses based on the regulated entities' track record, risk assessment and effectiveness of different options. The gradation of available sanctions must be adequate to allow credible deterrence through the escalation of sanctions.

From a behavioural perspective, a sanctioning led-approach has shown ineffective at deterring poor behaviour and is premised on the faulty assumption that everyone is likely to misbehave. In fact, research shows that only a small number of people intentionally do bad things. Research further shows that most people want to the do the right thing most of the time but they might not know what or how to do it. Therefore, what is needed is help to do the right thing (Hodges, 2016).

Taking this into consideration, the Scottish Environment Protection Agency produced a range of regulated entity compliance categories, along with the engagement approach to use to encourage compliance for each category (Figure 3). The compliance spectrum is a key concept within SEPA's regulatory strategy, *One Planet Prosperity* (SEPA, 2016[4]).

Figure 3. Compliance and engagement spectrum

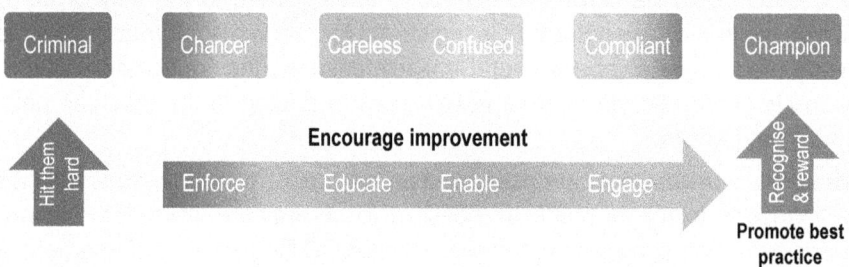

Source: (Hodges, 2016[5]), Regulatory Powers and Enforcement, PowerPoint presentation, University of Oxford, https://bit.ly/2LsbD2P (accessed 18 July 2019); (SEPA, 2016[4]), *One Planet Prosperity – Our Regulatory Strategy*, https://www.sepa.org.uk/media/219427/one-planet-prosperity-our-regulatory-strategy.pdf.

Output and outcome

Data collection

The EPA collects data from regulated entities in the framework of its licensing and enforcement activities, and recent innovations in processes have made important efficiency gains. The EPA recently introduced an electronic system for environmental and radiological protection licensing, monitoring and reporting ("LEMA") that centralises data requests and has reduced the burden on regulated entities. The EPA also collects large quantities of data to monitor and assess Ireland's environment, fulfilling several statutory reporting duties to the national government and the EU (e.g. water quality monitoring for the WFD).

A newly-introduced data analytics team provides opportunities for targeted interventions. In 2018, the EPA established a small data analytics team to pilot the use of data science, spatial analysis, earth observation and data visualisation techniques, working in close collaboration with EPA subject matter experts. For example, working with the urban wastewater treatment data that Irish Water submits, the analytics team produced an urban wastewater scorecard that allows inspectors to quickly focus on the specific plants and parameters that are a problem among the thousands of data points.

Monitoring and reporting on performance

Regulated entities

The EPA implements a transparent reporting system on the performance of regulated entities in all areas of its work, although information is not always easy to find. An annual review of the performance of facilities is carried out and published in the annual reports on drinking water, wastewater and industrial and waste licence enforcement, along with Annual Environmental Reports (AER) which are submitted by all licensees and published online. Reports can be difficult to find on the website.

Ireland's environment

The EPA monitors, assesses and reports on Ireland's environment, but information is often dispersed and difficult to navigate online. The EPA makes data accessible to the public through its reports, such as the State of the Environment report published every four years, annual reports on drinking water, urban wastewater, bathing water, water quality, air quality, greenhouse gas emission inventories and projections and waste. Some data is reported in near real-time (for example, air quality, hydrometrics). The reports, data and indicators on environmental performance in various sectors can be found on the EPA website or on specialised websites hosted by the EPA (such as catchments.ie and beaches.ie), but the information is dispersed and difficult to navigate. For example, the "Ireland's Environment" pages (http://www.epa.ie/irelandsenvironment/) present some data on waste, whereas the National Waste Statistics pages (http://www.epa.ie/nationalwastestatistics/) present more extensive and detailed information, yet the Ireland's Environment page does not state that more detailed data is available elsewhere on the site. Data are also published on the national open data portal (https://data.gov.ie/organization/environmental-protection-agency) and Ireland's Environmental Open Data Portal (https://data.epa.ie/). Although publically available, many EPA-published datasets are difficult to use and understand for the general public or non-specialist audiences. The Environmental Open Data Portal, for example, is primarily intended as a resource for software developers.

The EPA

The EPA operates in the framework of a strategic plan 2016-2020, *Our Environment, Our Wellbeing*, which sets out five goals and fifteen associated outcomes (three per goal), but the lack of measurable targets means that it cannot easily be used to monitor the performance of the EPA. Each outcome encompasses several objectives. Measurement and performance monitoring is not possible as the plan does not include targets or baselines. The plan is translated into annual work programmes that list tasks for each Office. Some tasks are also defined as key performance indicators (KPIs). A progress report on the work programme is presented to the Board once a month. This monitoring focuses on the implementation of activities and projects, rather than the quality of processes and ultimately, the performance and impact of EPA.

Parallel processes for monitoring, with different sets of indicators, are burdensome and unsuited as a tool to drive improvement. Internally, there are two layers of reporting to the Board. First, the monthly monitoring on the implementation of annual work programme, as described above. Second, each Office submits narrative reports to the Board three times per year. These reports are to be commended as they include indicators that go beyond outputs to include indicators of sector performance (e.g. number of boil water notices issued) and the quality of regulatory processes (e.g. handling of complaints, results of legal action). However, it is not clear whether there is consistent reporting on the same indicators over time. There do not appear to be any targets associated with these indicators. Separate reporting by Office may obscure visibility of EPA's overall performance. Externally, a third layer is added through the Performance Delivery Agreement (PDA) with DCCAE and DHPLG. The PDA defines a small number of key high-level metrics and indicators that EPA must report on (Table 5) as well as over a hundred other performance indicators that are not reported on. These indicators are not explicitly linked to the strategic goals of the EPA. Performance delivery monitoring is carried out twice a year, at mid-year and year end.

Table 5. Eight high-level metrics and indicators included in the PDA 2018

Metrics/indicator
Number of Environmental and Radiological Decisions
Number of Industrial/Waste site visits
Number of Urban Wastewater and Drinking Water Site Visits
Number of EPA Reports published
Number of Reports on Environmental Research Projects published
Number of Open Data datasets on the DPER Open Portal
Number of visits to EPA website
Number of environmental queries from the public answered

Overall, indicators focus on outputs rather than more meaningful indicators of outcomes and performance. The KPIs included in the annual work programme and those defined under the PDA tend to focus on metrics (outputs) rather than quality of processes (e.g. time to process licence applications), outcomes of activities, or overall sector performance (i.e. water or air quality, safety of industrial sites...) (Box 8). The performance measurement indicators listed in the PDA tend to be vague ("legislative obligations met", "plan implemented"...). The KPIs in the annual work programme used for internal monitoring also typically focus on outputs, such as the approval of plans and strategies by the Board rather than indicators that can be used to monitor performance (Box 9). Furthermore, internal performance monitoring frameworks do not include quantitative targets.

Box 8. Measuring organisational and policy performance: the Canada Energy Regulator's departmental results framework (Canada)

The Canada Energy Regulator (CER) measures its effectiveness in delivering its mandate using a Departmental Results Framework (DRF). Within the DRF, the CER links its core responsibilities with outcomes, to which it attaches indicators that seek to demonstrate its performance in delivering its mandate. The DRF provides information that the CER uses to refine the approach that it takes to delivering its mandate over time.

For each core responsibility, the CER aggregates specific activities under a program to which the outcomes that the CER is seeking to achieve are linked to a performance indicator and target, along with the intent of the measure.

> The CER has also established a Performance Measurement Evaluation Committee (PMEC). The PMEC, composed of senior CER officials and its CEO, reviews the DRF and presents the results to the CER'S Board of Directors quarterly. In the DRF quarterly performance report, the results and actions that the CER proposes to undertake in light of its performance are determined.
>
> Source: Information provided by the Canada Energy Regulator (CER), December 2019.

> **Box 9. Key Performance Indicators**
>
> Key performance indicators (KPIs) provide a means to measure whether organisations are performing in relation to their strategic goals and objectives. A manageable number of well-designed KPIs give a clear picture of current levels of performance and can aid decision-making. Each KPI should be clearly linked to a strategic objective and accompanied by a target or benchmark.
>
> Indicators of output from regulatory activity capture whether regulatory decisions, actions and interventions are effective (e.g. decisions taken which were upheld). Indicators of direct outcomes or the impact of outcomes could include, for example, compliance with the regulator's decisions.
>
> Indicators of wider outcomes ("watchtower" indicators) can be included as learning (rather than accountability) indicators. These could include, for example, service and infrastructure quality (e.g. frequency and reliability of services to consumers).
>
> Notes: The framework for performance indicators was proposed in the initial methodology for the performance assessment framework for economic regulators (PAFER) discussed with the OECD Network of Economic Regulators (NER). It has been refined to reflect feedback from NER members and the experience of other regulators in assessing their own performance.
>
> Source: (OECD, 2015[6]), Driving Performance at Colombia's Communications Regulator, Figure 3.3, Paris, http://dx.doi.org/10.1787/9789264232945-en.

Formally, the EPA is accountable to the House of Oireachtas but lacks structured engagement mechanisms to report on its performance. The EPA prepares an annual report and account that are laid before the House of Oireachtas by the Minister of DCCAE, and published online by the EPA when approved by the Oireachtas. The report is structured in line with four out of five of the goals of the EPA strategic plan – regulation, knowledge, advocacy and "organisationally excellent" – while omitting "responding to key environmental challenges". The EPA is often called to appear before Joint Oireachtas Committees to discuss particular issues, or to submit written answers to parliamentary questions, but the EPA does not engage in a structured dialogue with the legislature on its performance or the findings of the annual report.

Recommendations

- **Share the experience of introducing a data analytics team with other regulators nationally and internationally**. The EPA should also continue to explore how data analytics can be applied to its work to improve its regulatory activities and knowledge functions.
- **Improve the accessibility of information on the website(s).** The EPA produces a wealth of useful technical information but does not make it readily accessible. The website needs to be streamlined and entirely redesigned to make data and information more accessible. Performance

reports on regulated entities need to be found easily. Data and information on Ireland's environment could be centralised and presented in formats that are easy to understand, download and use. Data visualisation tools could be used to help the public and non-technical audiences understand the data.

- **Better engage with regulated entities around performance.** For example, the EPA could organise events by sector (e.g. waste) or type of regulated entity (e.g. intensive farming) to discuss performance, identify best practices and recognise 'champions'. Such events could become important drivers of compliance and build trust between the regulator and industry.
- **Develop a unified, outcome-based system for EPA performance assessment and reporting.** The EPA should define a manageable number of KPIs with time-bound targets that capture the quality of processes (e.g. time to process licence applications), the outcomes of activities, and the overall sector performance (e.g. safety of industrial sites...). It will be important to agree on the KPIs and associated targets in partnership with the parent Departments so that they become a useful tool to focus the dialogue around the executive's expectations for the EPA. Finally, reporting could be centralised through the corporate governance programme in OCCS, rather than by Office, to enable a holistic view of the EPA's overall performance.
- **Consider including indicators of wider environmental outcomes (e.g. air quality, water quality…) in performance evaluation frameworks.** These indicators can be a "watchtower" to loop back and help identify problem areas, orient decisions and identify priorities. They should be used as learning rather than accountability indicators, recognising that EPA is not solely responsible or accountable for these outcomes.
- **Invest in outward, results-based communications to demonstrate the impact of EPA activities to a number of stakeholders.** The EPA should put in place a regular engagement activity with the Oireachtas to increase accountability as well as share the EPA's role and activities with the legislature. This could, for example, take the form of a yearly event. In all its communications around performance, the EPA must use plain language suited to a non-technical audience.

References

DPER (2016), *Code of Practice for the Governance of State Bodies*, Department of Public Expenditure and Reform, Ireland, https://govacc.per.gov.ie/wp-content/uploads/Code-of-Practice-for-the-Governance-of-State-Bodies.pdf. [1]

EEB (2017), *Burning the evidence: A case study on large combustion plants*, https://eeb.org/library/burning-the-evidence-a-case-study-on-large-combustion-plants/. [3]

Hodges, C. (2016), *Regulatory Powers and Enforcement*, PowerPoint Presentation, University of Oxford, https://bit.ly/2LsbD2P (accessed on 15 November 2019). [5]

OECD (2017), *Driving Performance at Mexico's National Hydrocarbons Commission*, The Governance of Regulators, OECD Publishing, Paris, https://dx.doi.org/10.1787/9789264280748-en. [2]

OECD (2015), *Driving Performance at Colombia's Communications Regulator*, OECD Publishing, Paris, https://dx.doi.org/10.1787/9789264232945-en. [6]

SEPA (2016), *One Planet Prosperity – Our Regulatory Strategy*, https://www.sepa.org.uk/media/219427/one-planet-prosperity-our-regulatory-strategy.pdf (accessed on 20 December 2019). [4]

1 Regulatory and sector context

This chapter provides an overview of Ireland's public institutions and describes the main features of the regulatory framework that determines the functions of Ireland's Environmental Protection Agency (EPA).

Ireland is a parliamentary republic with a bicameral parliament, an executive branch headed by a prime minister and a directly elected president (Figure 1.1). Ireland has 31 municipalities, comprising 26 county councils, three city councils, and two city and county councils (OECD, 2016[1]).

Figure 1.1. Ireland's public institutions

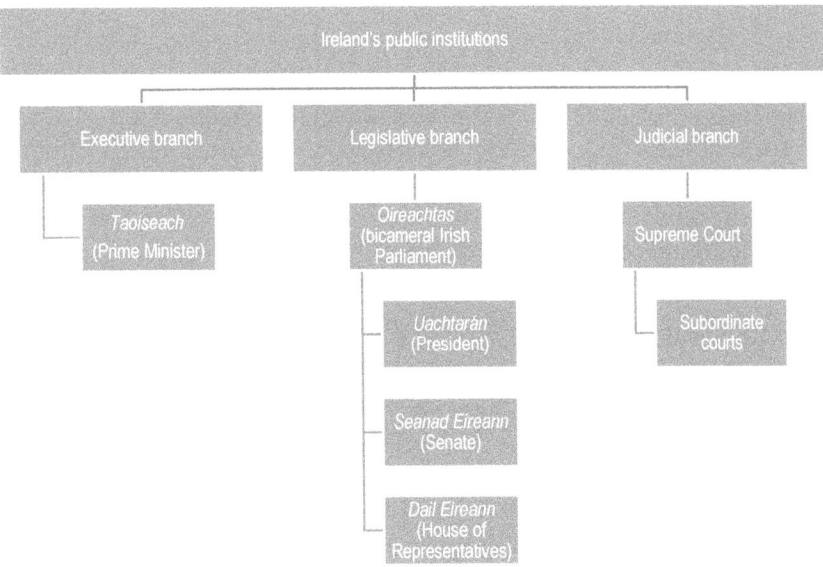

Source: Author's elaboration.

Environmental policy and regulations in Ireland

The majority of environmental policy in Ireland is led primarily by the Department for Communications, Climate Action and Environment, in conjunction with the Department for Housing, Planning and Local Government that has responsibility for water. Ireland's environmental policy is also significantly shaped by its membership of the European Union (Department of Communications, Climate Action and Environment, 2019[2]). The following section describes the main features of the areas where the EPA has responsibilities for environmental protection.

Integrated pollution and industrial activities licensing

The Environmental Protection Agency Act of 1992 (as amended) governs EPA's licensing activities and creates an obligation to regulate certain activities through an integrated pollution control licence. The EPA licence covers emissions to air, water and land from a facility as well as its environmental management (Environmental Protection Agency, n.d.[3]). Among other requirements, an applicant must demonstrate and the Agency must be satisfied that the emissions from the activity will not cause significant environmental pollution in order to receive a licence. The activities within the scope of licensing are listed in the First Schedule to the Environmental Protection Agency Act (as amended) and the Third and Fourth Schedules of the Waste Management Act 1996 (as amended) ((n.a.), 1992[4]; (n.a.), 1996[5]).

The 2003 Protection of the Environment Act aligned the licensing regimes with the requirements in the EU Directive on Integrated Pollution Prevention and Control (Council Directive 96/61/EC), broadening the range of activities covered under the permitting regime. The 2003 Act strengthened the requirement that activities must meet the standard of the Best Available Technology (BAT) (OECD, 2010[6]). The Environmental Protection Agency (Integrated Pollution Control) (Licensing) Regulations of 2013 present additional regulations within the meaning of Part IV of the amended Act of 1992 that established the Integrated Pollution Control licensing regime ((n.a.), 2013[7]).

The EPA is also the competent authority for regulations under the Industrial Emission Directive 2010/75/EU. The Environmental Protection Agency (Industrial Emissions) (Licensing) Regulations of 2013 impose a licensing requirement and additional rules upon industrial activities (Department of Communications, Climate Action and Environment, n.d.[8]; (n.a.), 2013[9]; (n.a.), 1992[4]).

Water

The Water Framework Directive 2000/60/EC defines the overarching arrangements governing water quality in EU Member States, setting the objective of attaining non-deterioration of water status and good status for all EU waters ((n.a.), 2000[10]). Its daughter directives – the Groundwater Directive (2006/118/EC) and the Environmental Quality Standards Directive (2008/105/EC) – contain measures to limit groundwater pollution and establish standards for certain pollutants in surface water (European Environment Agency, n.d.[11]). Ireland transposed the Directive within its Water Policy Regulations (S.I. No. 722 of 2003), Surface Waters Regulations (S.I. No. 272 of 2009) and Groundwater Regulations (S.I. No. 9 of 2010).

The European Union (Water Policy) Regulations 2014 marked a reform of water governance in Ireland, shifting the implementation approach for the Water Framework Directive. The 2014 Regulations established a Water Policy Advisory Committee to advise the Minister for the Environment, Community and Local Government on certain aspects of water policy and promote activities to support the implementation of the Water Framework Directive. It transferred certain local authority duties to the EPA and the Minister for the Environment, Community and Local Government (now the Minister for Housing, Planning and Local Government) ((n.a.), 2014[12]). The National River Basin Management Plan 2018-2021 is based on the integrated catchment management approach and embeds these reforms in the implementation structures and roles (Department of Housing, Planning and Local Government, 2018[13]).

A suite of European Directives provide the backdrop for regulation of Ireland's water and wastewater sector in tandem with the Water Framework Directive, including the Urban Waste Water Treatment Directive (91/271/EEC) and the Drinking Water Directive (98/83/EC). The Urban Waste Water Treatment Directive defines standards for the collection, treatment and disposal of wastewater and establishes monitoring requirements for discharges from urban areas (European Environment Agency, 2018[14]). The Drinking Water Directive establishes drinking water quality standards for 48 microbiological, chemical and indicator parameters (European Commission, 2018[15]). Together, the three directives form the backbone of Europe's water policy.

Recent years have seen significant changes in the structure and legal framework of the Irish water and wastewater sector. The Water Services Act 2013 established a new national water utility, Irish Water, that unified the water and wastewater services of 31 local authorities under one provider (Irish Water, n.d.[16]). In parallel, the Commission for Regulation of Utilities (then the Commission for Energy Regulation) became the economic regulator of the public water and wastewater sector (OECD, 2018[17]). The Drinking Water Regulations (S.I. No. 122 of 2014) as amended give EPA supervisory powers over public water supplies (Environmental Protection Agency, n.d.[18]). EPA provides authorisation for wastewater discharges from Water Services Authorities, satisfying provisions in a number of EU Directives on discharge of hazardous substances and wastewater pollution reduction (Environmental Protection Agency, n.d.[19]).

In 2018, the Irish Government released a water services policy statement expressing expectations for the delivery and development of water and wastewater services through 2025. This plan, which fulfils a statutory obligation under the Water Services Act 2017 (amending the Water Services Acts 2007, 2013 and 2014), serves as the scaffolding within which Irish Water's funding and investment plans will be developed and informs the regulation of the company. In addition, it guides the development of rural water services alongside a review of these services that started in April 2018 (Department of Housing, Planning and Local Government, 2018[13]).

Waste

The Waste Framework Directive (Directive 2008/98/EC) sets the overarching framework for European waste policy. The Waste Framework Directive establishes rules for waste management in EU Member States based on the principles of waste hierarchy, polluter pays and extended producer responsibility (European Commission, 2016[20]). The Irish Waste Management Act 1996 and the European Communities (Waste Directive) Regulations 2011 implement the provisions of the Directive on a national level (Department of the Environment, Community and Local Government, 2012[21]). Several additional EU Directives influence Ireland's waste policy, including the Packaging Directive (94/62/EC); Waste Electrical and Electronic Equipment (WEEE) Directive (2002/96/EC); Restriction of Hazardous Substances in WEEE Directive (2002/95/EC); End of Life Vehicles (ELV) Directive (2000/53/EC); Batteries Directive (2006/66/EC); and Landfill Directive (1999/31/EC) (Department of the Environment, Community and Local Government, 2012[21]).

The government has presented its waste management policy in a set of four policy documents produced since 1998, with the current policy *"A Resource Opportunity"* launched in 2012. The policy outlines measures to further progress towards becoming a "recycling society", focusing on resource efficiency and reduction of landfill disposal of municipal waste (Department of the Environment, Community and Local Government, 2012[21]). In 2016, responsibility for waste management policy shifted to the Department of Communications, Climate Action and Environment (Citizens Information, 2016[22]). Ireland's National Waste Prevention Programme, delivered by the EPA and overseen by the National Waste Prevention Committee, provides guidance and support for businesses, households and the public sector to become more resource efficient (Department of Communications, Climate Action and Environment, n.d.[23]). Policies and programmes on the national level guide the waste management actions of local and regional bodies.

The responsibility for waste management planning falls to local authorities under Part II of the Waste Management Act, 1996. A regional waste management plan is produced for three regions – Connacht-Ulster, Southern and Eastern-Midlands (Environmental Protection Agency, n.d.[24]). The creation of three Waste Enforcement Regional Lead Authorities (WERLAs) in 2015 marked another significant step towards the regionalisation of Ireland's approach to waste enforcement. Cork County Council, Dublin City Council and the combined Leitrim and Donegal Council Councils serve as the three WERLAs, one for each of the regions covered under the regional waste management plans. WERLAs co-ordinate enforcement and establish common objectives and priorities, while local authorities maintain a "first responder" role for waste violations (Department of Communications, Climate Action and Environment, n.d.[25]).

Dumping at sea is regulated under a separate regime. The Dumping at Sea Acts 1996 to 2012 execute requirements on ocean dumping in the London Convention 1972 and the OSPAR Convention 1992 (Environmental Protection Agency, 2012[26]). Amendments in 2010 to the Dumping at Sea Act 1996 gave EPA the function of issuing Dumping at Sea Permits (Environmental Protection Agency, n.d.[27]). Sea disposal of dredged material and inert material of natural origin requires a permit (Environmental Protection Agency, 2012[26]).

Air

Ireland is currently developing its first National Clean Air Strategy, which will create a framework to facilitate cross-government policies for clean air (Department of Communications, Climate Action and Environment, 2017[28]). The strategy is a response to several recent developments, including revised World Health Organization estimates of public health risks from air pollution and EU Clean Air Package legislation finalised in recent years. The strategy will be developed against an evolving background of EU and national legislation.

A range of EU Directives aim to reduce emissions and improve ambient air quality, driven by the Clean Air Policy Package and its objectives within the Clean Air for Europe (CAFE) Programme (European Commission, 2019[29]). The Ambient Air Quality Directive, its daughter directives and the National Emission Ceilings Directives provide the framework for EU air policy (European Commission, 2019[29]). The Ambient Air Quality Directive (2008/50/EC) replaced the Air Quality Framework Directive and three of its daughter directives. The fourth daughter directive to the Ambient Air Quality Directive remains in force. Together, these directives set legally-binding limits for concentrations of priority pollutants in the ambient air (Department of Communications, Climate Action and Environment, 2017[28]). The National Emissions Ceiling Directive (Directive 2016/2284/EU) commits member nations to reductions of five pollutants by 2030: sulphur dioxide, nitrogen oxides, volatile organic compounds, ammonia and fine particulate matter. Alongside these directives, EU product standards and sector-specific pollution control measures have also played a role in emissions reduction.

In Ireland, the Air Pollution Act 1987 serves as a foundation for later national legislation on air pollution. As amended, the Act assigns authorities the responsibility to regulate emissions of pollutants from certain sources – local authorities regulate small sources while the EPA regulates larger sources. Subsequent regulations have served to transpose requirements from EU directives into national law. Ireland transposed the CAFE directive in their Air Quality Standards Regulations 2011 (S.I. No. 180 of 2011). Ireland transposed the fourth daughter directive in the Arsenic, Cadmium, Mercury, Nickel and Polycyclic Aromatic Hydrocarbons in Ambient Air Regulations 2009 (S.I. No. 58 of 2009) (Environmental Protection Agency, n.d.[30]). Ireland's European Union (National Emission Ceilings) Regulations 2018 (S.I. No. 232 of 2018) transpose the requirements under the National Emissions Ceiling Directive. Under these regulations, EPA must prepare an annual inventory and projections of emissions of the five pollutants. EPA issues VOC permits under Environmental Protection Agency Act, 1992 (Control of volatile organic compound emissions resulting from the storage of petrol and its distribution) Regulations, 1997 (S.I. No. 374 of 1997) (Environmental Protection Agency, n.d.[31]).

European and national legislation provide a framework for the monitoring and control of ozone-depleting substances for Ireland, a party to the Montreal Protocol since 1988 ((n.a.), 2019[32]). Following the European Regulation on Substances that Deplete the Ozone Layer (Regulation (EC) No. 2037/2000), Ireland's Control of Substances that Deplete the Ozone Layer Regulations 2006 (S.I. No. 281 of 2006) introduced regulations to control ozone-depleting substances. The EPA is the competent authority under this regulation, and it has responsibilities that include receiving licenses, export authorisations, and reports as well as conducting investigations and random checks on imports of controlled substances ((n.a.), 2006[33]). The EU Ozone Regulation (Regulation (EC) No 1005/2009) increases ambition by, for example, prohibiting the use of certain substances (European Commission, n.d.[34]). This regulation was followed in Ireland by the Control of Substances that Deplete the Ozone Layer Regulations 2011 (S.I. No. 465 of 2011) (Environmental Protection Agency, 2011[35]).

With the global phase-out of ozone-depleting substances catalysed by the Montreal Protocol, the use of fluorinated gases as replacements has increased. While not ozone-depleting substances, these gases have a significant global warming potential. The European Regulation on fluorinated greenhouse gases (Regulation (EU) No. 517/2014) imposed new measures to reduce fluorinated gases by phasing-down allowed sales of HFCs in the EU market and banning fluorinated gases with high global warming potential ((n.a.), 2014[36]). Ireland's own European Union (Fluorinated Greenhouse Gas) Regulations 2016 (S.I. No. 658 of 2016) designate the EPA as the competent authority ((n.a.), 2016[37]).

Climate change and emissions trading

In 2016, Ireland convened a Citizens' Assembly, a randomly-selected but representative group of citizens, to deliberate on a range of issues including climate change. One of the questions considered by the assembly was "how can the state make Ireland a leader in tackling climate change?" (The Citizens'

Assembly, 2017[38]). The assembly resulted in a final report and a series of recommendations, providing a strong foundation for the government's Climate Action Plan, released in June 2019 (Government of Ireland, 2019[39]). The Climate Action Plan charts a course towards decarbonisation through measures such as instituting carbon-proofing policies, establishing carbon budgets, strengthening the Climate Change Advisory Council and increasing accountability to the Oireachtas (Department of Communications, Climate Action and Environment, 2019[2]).

The Climate Action Plan includes Ireland's target to achieve net-zero emissions by 2050 (Government of Ireland, 2019[39]). This goal builds upon the statutory basis provided in the Climate Action and Low Carbon Development Act 2015 (Department of Communications, Climate Action and Environment, n.d.[40]). Intermediate targets support this longer-term goal. The EU's nationally determined contribution (NDC) under the Paris Agreement is to reduce greenhouse gas emissions by at least 40% by 2030 compared to 1990. The target will be delivered collectively by the EU with reductions in the Emissions Trading Scheme (ETS) and non-ETS sectors amounting to 43% and 30% by 2030 compared to 2005 respectively.

In relation to the ETS, established in Directive 2003/87/EC and amendments, it is the cornerstone of the EU's approach to tackling climate change. It covers emissions from electricity generation and large industry. EPA is the competent authority for the implementation of the EU ETS in Ireland. One-hundred and one stationary installations are engaged in activities listed in Annex 1 of Directive 2003/87 and amendments in Ireland, and are regulated by Greenhouse Gas Emission Permits. In addition, fourteen aviation operators are also included in the ETS (Environmental Protection Agency, 2019[41]).

There is a single, EU-wide cap on emissions under the EU ETS and an agreed limit to reduce greenhouse gas emissions by 21% compared with 2005. Under Phase IV, which will run from 2021-2030, the sectors covered by EU ETS must reduce their emissions by 43% by 2030 compared to 2005 levels. In relation to those sectors that fall outside the EU ETS - non-ETS sector emissions – Ireland has a binding target of 20% reduction for non-ETS sector emissions by 2020 established in the EU Effort Sharing Decision (compared to a 2005 baseline). The Effort Sharing Regulation translates the commitments made as part of the Paris Agreement into binding annual greenhouse gas emission targets for each Member State for the period 2021-2030. Under the Effort Sharing Regulations, Ireland has a greenhouse-gas reduction target of 30% emissions reduction between 2021 and 2030 (compared to a 2005 baseline) (Department of Communications, Climate Action and Environment, n.d.[42]).

Radiation

The EU has established binding requirements for the use of ionising radiation in Member States in the Basic Safety Standards Directive. This Directive has been implemented in Ireland through the Ionising Radiation Regulations (S.I. No. 30 of 2019) which relate to workers and members of the public and the Medical Exposure Regulations (S.I. No. 256 of 2018) which cover patient protections.

EPA is the responsible authority for regulations to protect against occupational and public exposures while the Health Information and Quality Authority administer regulation relating to patient exposures. The EPA monitors radiation levels and assesses public exposure, helps create radiological emergency plans, follows international developments related to radiological and nuclear safety, and managing certain radiation protection services (Environmental Protection Agency, 2015[43]).

New regulations on non-ionising radiation (NIR) were published in May 2019, giving additional functions to the EPA. Under the Radiological Protection Act 1991 (Non-Ionising Radiation) Order 2019 (S.I. No. 190 of 2019), the EPA has responsibility to provide advice to the Minister for Communications, Climate Action and Environment on public exposure to NIR fields, including advice on relevant standards; to provide general information to other Ministers of Government, local authorities and members of the public on public exposure to NIR; to monitor scientific, technological and other developments on matters pertaining to public exposure to NIR; and to monitor the public exposure to NIR to support the EPA advisory role.

Genetically modified organisms

The European Union's legal framework for GMOs introduces a safety assessment before GMOs can be placed on the market, harmonises procedures for risk assessment and authorisation, requires labelling of GMOs on the market, and imposes requirements to ensure the traceability of GMOs on the market (European Commission, n.d.[44]). It is built upon the building blocks of (1) Directive 2001/18/EC on the deliberate release of GMOs into the environment, (2) Regulation (EC) 1829/2003 on genetically modified food and feed, (3) Regulation (EC) No. 1830/2003 concerning the traceability and labelling of genetically modified organisms and the traceability of food and feed products produced from genetically modified organisms and amending Directive 2001/18/EC, (4) Directive (EU) 2015/412 amending Directive 2001/18/EC as regards the possibility for the Member States to restrict or prohibit the cultivation of GMOs in their territory, (5) Regulation (EC) 1830/2003 concerning the traceability and labelling of genetically modified organisms and the traceability of food and feed products produced from genetically modified organisms, (6) Directive (EU) 2018/350 amending Directive 2001/18/EC as regards the environmental risk assessment of genetically modified organisms (7) Directive 2009/41/EC on contained use of genetically modified micro-organisms and (8) Regulation (EC) 1946/2003 on transboundary movements of GMOs (European Commission, n.d.[44]).

The requirements of these directives are reflected in Irish national legislation. The national legal framework for GMOs is based on the Genetically Modified Organisms (Deliberate Release) Regulations 2003 (S.I. No. 500 of 2003), the Genetically Modified (Contained Use) Regulations 2001-2010, and the Genetically Modified Organisms (Transboundary Movement) Regulations 2004 (S.I. No. 54 of 2004). EPA serves as the competent authority for the relevant Directives and national GMO regulations (Department of Communications, Climate Action and Environment, n.d.[45]).

References

(n.a.) (2019), "United Nations Treaty Collection", https://treaties.un.org/Pages/ViewDetails.aspx?src=TREATY&mtdsg_no=XXVII-2-a&chapter=27&lang=en (accessed on 1 August 2019). [32]

(n.a.) (2016), *European Union (Fluorinated Greenhouse Gas) Regulations 2016*, Iris Oifigiúil, http://www.epa.ie/pubs/advice/air/fluorinatedgreenhousegases/F-gas_Regulations_SI_658_of_2016.pdf. [37]

(n.a.) (2014), *European Union (Water Policy) Regulations 2014*, Iris Oifigiúil, http://www.irishstatutebook.ie/eli/2014/si/350/made/en/print (accessed on 23 July 2019). [12]

(n.a.) (2014), *Regulation (EU) No 517/2014 of the European Parliament and of the Council of 16 April 2014 on fluorinated greenhouse gases and repealing Regulation (EC) No 842/2006*, Official Journal of the European Union, https://www.epa.ie/pubs/legislation/air/ods/Regulation 517 of 2014 Fgas.pdf (accessed on 24 July 2019). [36]

(n.a.) (2013), *Environmental Protection Agency Industrial Emissions Licensing Regulations 2013*, http://www.irishstatutebook.ie/eli/2013/si/137/made/en/pdf (accessed on 17 July 2019). [9]

(n.a.) (2013), *S.I. No. 283/2013 - Environmental Protection Agency (Integrated Pollution Control) (Licensing) Regulations 2013*, Irish Statute Book, http://www.irishstatutebook.ie/eli/2013/si/283/made/en/print (accessed on 1 August 2019). [7]

(n.a.) (2006), *Control of Substances that Deplete the Ozone Layer Regulations 2006*, Stationery Office, https://www.epa.ie/pubs/legislation/air/ods/SI%20281%20of%202006%20ODS%20Regs.pdf (accessed on 24 July 2019). [33]

(n.a.) (2000), *Directive 2000/60/EC of the European Parliament and of the Council of 23 October 2000 establishing a framework for Community action in the field of water policy*, https://eur-lex.europa.eu/resource.html?uri=cellar:5c835afb-2ec6-4577-bdf8-756d3d694eeb.0004.02/DOC_1&format=PDF (accessed on 23 July 2019). [10]

(n.a.) (1996), *Waste Management Act, 1996*, http://www.irishstatutebook.ie/eli/1996/act/10/enacted/en/print.html. [5]

(n.a.) (1992), *Environmental Protection Agency Act, as amended*, https://www.epa.ie/pubs/advice/process/First%20Schedule%20of%20EPA%20Act%201992%20as%20amended%20%20-V4%20Peat%20Regs%20updates.pdf (accessed on 17 July 2019). [4]

Citizens Information (2016), *Waste management legislation*, https://www.citizensinformation.ie/en/environment/waste_management_and_recycling/waste_management.html (accessed on 24 July 2019). [22]

Department of Communications, Climate Action and Environment (2019), *Climate Action Plan to Tackle Climate Breakdown*, https://www.dccae.gov.ie/en-ie/climate-action/topics/climate-action-plan/Pages/climate-action.aspx (accessed on 1 August 2019). [2]

Department of Communications, Climate Action and Environment (2017), *Cleaning Our Air Public Consultation to inform the development of a National Clean Air Strategy*, https://www.dccae.gov.ie/documents/Clean Air Strategy Public Consultation.pdf (accessed on 31 July 2019). [28]

Department of Communications, Climate Action and Environment (n.d.), *Climate Action and Low Carbon Development Act 2015*, https://www.dccae.gov.ie/en-ie/climate-action/legislation/Pages/Climate-Action-and-Low-Carbon-Development-Act-2015.aspx (accessed on 1 August 2019). [40]

Department of Communications, Climate Action and Environment (n.d.), *EU Emissions Targets*, https://www.dccae.gov.ie/en-ie/climate-action/topics/eu-and-international-climate-action/2020-eu-targets/Pages/default.aspx (accessed on 1 August 2019). [42]

Department of Communications, Climate Action and Environment (n.d.), *Genetically Modified Organisms*, https://www.dccae.gov.ie/en-ie/environment/topics/environmental-protection-and-awareness/genetically-modified-organisms/Pages/default.aspx (accessed on 24 July 2019). [45]

Department of Communications, Climate Action and Environment (n.d.), *Industrial Emissions*, https://www.dccae.gov.ie/en-ie/environment/topics/environmental-protection-and-awareness/industrial-emissions/Pages/default.aspx (accessed on 17 July 2019). [8]

Department of Communications, Climate Action and Environment (n.d.), *National Waste Prevention Programme (NWPP)*, https://www.dccae.gov.ie/en-ie/environment/topics/sustainable-development/waste-prevention-programme/Pages/default.aspx (accessed on 1 August 2019). [23]

Department of Communications, Climate Action and Environment (n.d.), *Regional Lead Authorities*, https://www.dccae.gov.ie/en-ie/environment/topics/waste/enforcement/enforcement-structures/Pages/Waste-Enforcement-Regional-Lead-Authorities.aspx (accessed on 24 July 2019). [25]

Department of Housing, Planning and Local Government (2018), *Water Services Policy Statement*, https://www.housing.gov.ie/sites/default/files/publications/files/water_services_policy_statement_2018-2025.pdf (accessed on 30 July 2019). [13]

Department of the Environment, Community and Local Government (2012), *A Resource Opportunity Waste Management Policy in Ireland*, https://www.dccae.gov.ie/documents/A Resource Opportunity 2012 Small.pdf (accessed on 30 July 2019). [21]

Environmental Protection Agency (2019), *Questionnaire delivered to OECD*, http://On record with authors. [41]

Environmental Protection Agency (2015), *National Ambient Air Quality Monitoring Programme 2017-2022*, http://www.epa.ie (accessed on 1 August 2019). [43]

Environmental Protection Agency (2012), *Dumping at Sea EPA Enforcement and Permitting Booklet*, https://www.epa.ie/pubs/advice/dumping%20at%20sea/EPA%20Dumping%20at%20Sea_web.pdf (accessed on 24 July 2019). [26]

Environmental Protection Agency (2011), "Control of Substances that Deplete the Ozone Layer Regulations 2011 (S.I. No. 465 of 2011)", *Environmental Protection Agency (EPA)*, https://www.epa.ie/pubs/advice/air/ods/irishodsregulations.html (accessed on 1 August 2019). [35]

Environmental Protection Agency (n.d.), "Air quality standards", *Environmental Protection Agency*, https://www.epa.ie/air/quality/standards/ (accessed on 31 July 2019). [30]

Environmental Protection Agency (n.d.), *Drinking Water*, Environmental Protection Agency (EPA), https://www.epa.ie/water/dw/ (accessed on 23 July 2019). [18]

Environmental Protection Agency (n.d.), *Integrated Pollution Control (IPC) Licensing*, https://www.epa.ie/licensing/ipc/ (accessed on 17 July 2019). [3]

Environmental Protection Agency (n.d.), *Licensing - Dumping at Sea (DaS permits)*, Environmental Protection Agency (EPA), https://www.epa.ie/licensing/watwaste/dumping/ (accessed on 18 July 2019). [27]

Environmental Protection Agency (n.d.), *Licensing - Petrol Storage & Distribution (VOC permits)*, Environmental Protection Agency (EPA), https://www.epa.ie/licensing/air/voc/ (accessed on 23 July 2019). [31]

Environmental Protection Agency (n.d.), *Regional waste management plans*, Environmental Protection Agency (EPA), https://www.epa.ie/waste/policy/regional/ (accessed on 18 July 2019). [24]

Environmental Protection Agency (n.d.), *Waste Water Discharge Authorisation*, Environmental Protection Agency (EPA), https://www.epa.ie/licensing/watwaste/wwda/ (accessed on 18 July 2019). [19]

European Commission (2019), *Reduction of National Emissions*, https://ec.europa.eu/environment/air/reduction/index.htm (accessed on 31 July 2019). [29]

European Commission (2018), *Drinking water legislation*, https://ec.europa.eu/environment/water/water-drink/legislation_en.html (accessed on 30 July 2019). [15]

European Commission (2016), *Directive 2008/98/EC on waste (Waste Framework Directive) - Environment - European Commission*, https://ec.europa.eu/environment/waste/framework/ (accessed on 30 July 2019). [20]

European Commission (n.d.), *GMO legislation*, https://ec.europa.eu/food/plant/gmo/legislation_en (accessed on 1 August 2019). [44]

European Commission (n.d.), *Protection of the ozone layer*, https://ec.europa.eu/clima/policies/ozone_en (accessed on 1 August 2019). [34]

European Environment Agency (2018), *Urban Waste Water Treatment Directive*, https://www.eea.europa.eu/archived/archived-content-water-topic/water-pollution/prevention-strategies/urban-waste-water-treatment-directive (accessed on 30 July 2019). [14]

European Environment Agency (n.d.), *Legislative instrument details: Water Framework Directive (consolidated)*, Eionet: Reporting Obligations Database, https://rod.eionet.europa.eu/instruments/516 (accessed on 30 July 2019). [11]

Government of Ireland (2019), *Climate Action Plan 2019*, https://www.dccae.gov.ie/documents/Climate%20Action%20Plan%202019.pdf (accessed on 1 August 2019). [39]

Irish Water (n.d.), *About Irish Water*, https://www.water.ie/about-us/our-company/ (accessed on 23 July 2019). [16]

OECD (2018), *Driving Performance at Ireland's Commission for Regulation of Utilities*, The Governance of Regulators, OECD Publishing, Paris, https://dx.doi.org/10.1787/9789264190061-en. [17]

OECD (2016), *Ireland regional policy profile*, OECD, https://www.oecd.org/regional/regional-policy/profile-Ireland.pdf (accessed on 23 July 2019). [1]

OECD (2010), *OECD Environmental Performance Reviews – Ireland*, OECD Publishing, https://www.oecd-ilibrary.org/docserver/9789264079502-en.pdf?expires=1563877926&id=id&accname=ocid84004878&checksum=20C11C53362E2C24C388C4C46C1D3E0D (accessed on 23 July 2019). [6]

The Citizens' Assembly (2017), *How the State can make Ireland a leader in tackling climate change - The Citizens' Assembly*, https://www.citizensassembly.ie/en/How-the-State-can-make-Ireland-a-leader-in-tackling-climate-change/How-the-State-can-make-Ireland-a-leader-in-tackling-climate-change.html (accessed on 1 August 2019). [38]

2 Governance of Ireland's Environmental Protection Agency

The Performance Assessment Framework for Economic Regulators (PAFER) was developed by the OECD to help regulators assess their own performance. The PAFER structures the drivers of performance along an input-process-output-outcome framework. This chapter applies the framework to the governance of Ireland's Environmental Protection Agency (EPA) and reviews the existing features, the opportunities and challenges faced by the EPA.

Role and objectives

Ireland's Environmental Protection Agency (EPA) was established in 1993 as a public regulatory body with administrative and technical independence to protect and improve Ireland's environment. Over time, the EPA's functions have expanded beyond those originally set out in its founding statute – the Environmental Protection Agency Act, 1992 – in step with new regulations, EU directives and following the merger in 2014 with the Radiological Protection Institute of Ireland (Box 2.1).

Upon its creation the EPA took over responsibility for many functions previously carried out by local authorities. The EPA's founding was in part a response to the difficulty Ireland's local authorities faced in consistently implementing increasingly complex environmental legislation across the country. Local authorities still however retain significant responsibilities for environmental protection, and the EPA has a supervisory role in this regard.

The establishment of the EPA was also a political response to restore public trust in the state's environmental protection functions. Previously, local authorities were responsible for the potentially conflicting functions of local development and environmental protection. Faced with the growth of complex sectors, such as the pharmaceutical industry, the public had lost confidence in local authorities' abilities to protect the environment (Shipan, 2006[1]).

Box 2.1. Legislation

The main legal instruments pertaining to the EPA are:

- Environmental Protection Agency Act 1992
- Waste Management Act 1996
- Protection of the Environment Act 2003
- Radiological Protection Acts 1991 to 2014
- Amendments to the above Acts
- Regulations made under the above Acts and the European Communities Act

The full list of legislation is:

- Bathing Water Quality Regulations 2008
- Air Quality Standards Regulations 2011
- Carriage of Dangerous Goods by Road Act 1998
- Chemicals Act 2008 and 2010
- Containment of Nuclear Weapons Act 2003
- Control of Substances that Deplete the Ozone Layer Regulations 2011
- Dumping at Sea Acts 1996 to 2009
- Emissions of Volatile Organic Compounds from Organic Solvents Regulations 2002 (as amended)
- Environmental Protection Agency Act 1992 (as amended)
- Environmental Protection Agency Act 1992 (Control of Volatile Organic Compound Emissions resulting from Petrol Storage and Distribution) Regulations 1997
- European Communities (Water Policy) Regulations, 2003-2014

- European Communities (Environmental Assessment of Certain Plans and Programmes) Regulations 2004 to 2011
- Planning and Development (Strategic Environmental Assessment) Regulations 2004 to 2011
- European Communities (Access to Information on the Environment Regulations) 2007 to 2014
- European Communities (Birds and Natural Habitats) Regulations 2011 and 2013
- European Communities (Environmental Assessment of Certain Plans and Programmes) Regulations 2004 to 2011
- European Communities (Environmental Liability) Regulations 2008
- European Communities (Foodstuffs Treated with Ionising Radiation) Regulations, 2000
- European Communities (Good Agricultural Practices for the Protection of Waters) Regulations 2010
- European Communities (Greenhouse Gas Emissions Trading) (Aviation) Regulations 2010, as amended
- European Union (Batteries and Accumulators) Regulations 2014 as amended
- European Union (Drinking Water) Regulations 2014
- European Union (Fluorinated Greenhouse Gas) Regulations 2016
- European Union (Paints, Varnishes, Vehicle Refinishing Products and Activities) Regulations 2012
- European Union (Radioactive Substances in Drinking Water) Regulations 2016
- European Union (Waste Electrical and Electronic Equipment) Regulations 2014
- European Union (Fluorinated Greenhouse Gas) Regulations 2016
- Freedom of Information Act 2014
- Genetically Modified Organisms (Contained Use) Regulations 2001
- Genetically Modified Organisms (Deliberate Release) Regulation 2003
- Kyoto Protocol Flexible Mechanisms Regulations 2006
- Limitation of Emissions of Volatile Organic Compounds Due to the Use of Organic Solvents in Certain Paints, Varnishes and Vehicle Refinishing Products Regulations 2007
- Nuclear Test Ban Act 2008
- Persistent Organic Pollutants Regulations 2010
- Planning and Development Regulations 2001, as amended
- Radiological Protection Act, 1991 (as amended)
- Radiological Protection Act, 1991 (Ionising Radiation) Order, 2000
- Radiological Protection Act 1991 (Ionising Radiation) Regulations, 2019
- Radiological Protection Act 1991 (Non-Ionising Radiation) Order, 2019
- Waste Management (Certification of Historic Unlicensed Waste Disposal and Recovery Activity) Regulations 2008
- Waste Management (End-Of-Life Vehicles) Regulations 2006
- Waste Management (Facility Permit and Registration) Regulations 2007
- Waste Management (Hazardous Waste) Regulations 1998
- Waste Water Discharge (Authorisation) Regulations 2007
- Water Services Act 2007 (as amended)

Today EPA's licensing, permitting and enforcement activities cover waste, wastewater, industrial emissions (emissions to air, water and land, generation of waste, noise), greenhouse gases, contained use and controlled release of genetically modified organisms (GMOs), sources of ionising radiation, dumping at sea, and volatile organic compounds (VOCs). EPA is also an environmental authority for strategic environmental assessments.

Its monitoring, analysing and reporting functions span a broader range of environmental areas. These include air quality, water quality (rivers, lakes, bathing water, drinking water…), radiation levels, biodiversity, species and habitats (reporting only), greenhouse gases, waste generation and management, and land and soil.

In addition to its regulatory, monitoring and reporting functions, the EPA has statutory responsibility for co-ordinating and funding national research on environmental protection.

Functions

The EPA's functions can be divided into "regulation, knowledge and advocacy" (Figure 2.1).

Figure 2.1. EPA's functions

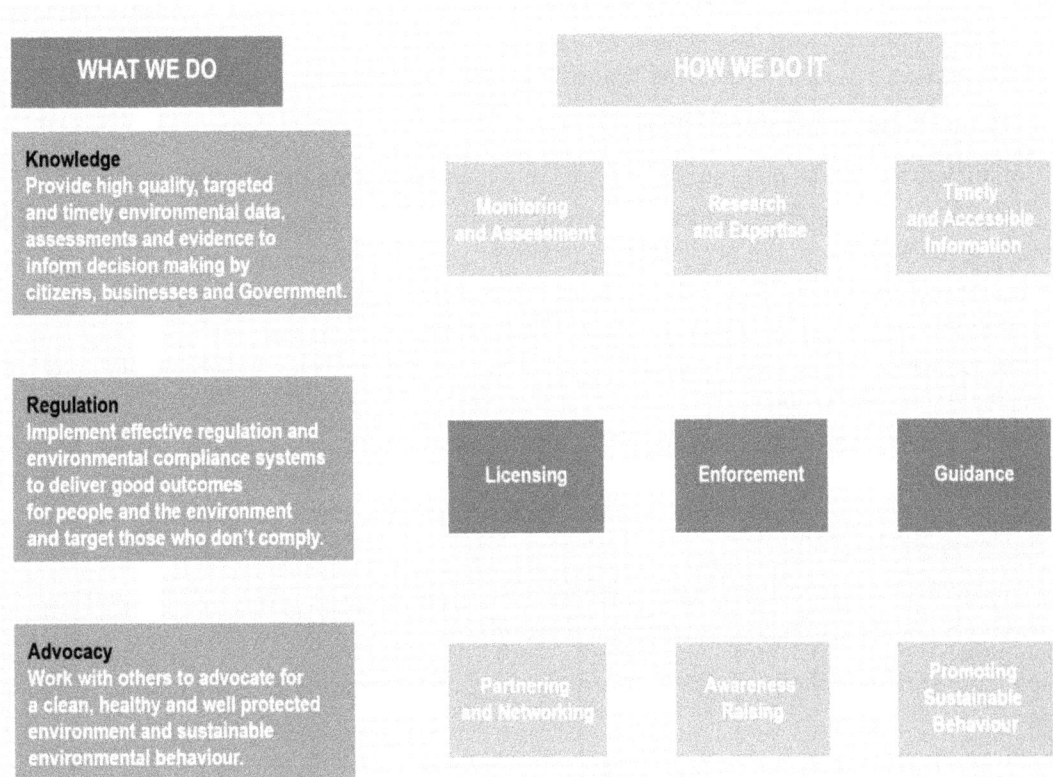

Source: (EPA, 2017[2]), EPA Annual Report and Accounts 2017, http://www.epa.ie/pubs/reports/other/corporate/EPA_AnnualReport_2017_EN_web.pdf.

Regulatory functions

Licensing, permitting, inspections and enforcement form the core of the EPA's regulatory work. The EPA's stated goal is "to implement effective regulation and environmental compliance systems to deliver good environmental outcomes and target those who don't comply."

EPA is responsible for licensing, permitting, consenting or certification of activities that could have an impact on the environment or on human health. EPA is responsible for the licensing and permitting of:

- Waste facilities
- Large-scale industrial activities
- Intensive agriculture
- Peat extraction
- The contained use and controlled release of GMOs
- Sources of ionising radiation (e.g. x-ray and radiotherapy equipment, industrial sources)
- Large petrol storage facilities (VOCs)
- Wastewater discharges
- Dumping at sea activities
- CO_2 emissions from large industrial facilities and the aviation sector through the EU Emissions Trading Scheme
- Medium combustion plants
- By-product and end-of-waste decisions

Additional licensing responsibilities are due to come into force in 2019. Ireland's River Basin Management Plan (2018-2021) commits to a new legislative framework for the management of water abstractions by 2019 that will include a requirement for the licensing of larger abstractions (>250m3/day) by the EPA.

EPA's enforcement responsibilities include:

- Conducting an annual programme of audits and inspections of EPA-licensed facilities.
- Supervising and reporting on local authority environmental performance, including investigating complaints from the public about local authorities.
- Supervising the supply of drinking water by public water suppliers. As well as auditing, inspecting and monitoring water services in Ireland, the EPA sets priorities for Irish Water to improve the national water infrastructure and uses its enforcement powers when these priorities are not being implemented satisfactorily.
- Working with local authorities and other agencies, including the National Waste Enforcement Steering Committee, to tackle environmental crime by co-ordinating a national enforcement network and targeting offenders and the Network for Ireland's Environmental Compliance and Enforcement (NIECE).
- Investigating failures to meet quality standards.
- Producing guidance on best practice for industry.

In addition to its supervisory role, the EPA has a statutory role to provide advice and assistance to local authorities.

The EPA is an environmental authority for strategic environmental assessments (SEAs) that assess the impact of plans and programmes on the Irish environment (e.g. major development plans, FoodWise2025, climate plans). The EPA has compiled guidance to help planning authorities when carrying out SEAs (EPA, 2018[3]). In certain situations the EPA carries out SEAs itself, for example, when it is leading the development of national plans (e.g. the National Hazardous Waste Management Plan). The EPA is also the competent authority for Environmental Impact Assessments for the activities it regulates.

The EPA is often responsible for developing national plans in relation to environmental protection, for example, the National Hazardous Waste Management Plan and the Persistent Organics Pollutions (POPs) Management Plan.

Knowledge functions

The EPA's stated goal is to "provide high-quality, targeted and timely environmental data, information and assessment to inform decision-making at all levels." The EPA is responsible for monitoring, assessment and reporting on a wide range of environmental outcomes and co-ordinates and funds a significant research programme to advance knowledge on environmental protection.

The EPA assesses and reports on:

- The State of Ireland's environment: the EPA must publish a State of the Environment Assessment every four years. The next report is due for publication in 2020.
- Water quality: rivers, lakes, transitional and coastal waters, groundwater, bathing water; drinking water.
- Hydrometrics: water resource and flows assessment and modelling.
- Catchments: to support river basin management planning and implementation.
- Air quality: ambient air quality monitoring, modelling and forecasting.
- Greenhouse gases: inventories and projections.
- Radiation: ambient radioactivity levels in air, foodstuffs and drinking water, marine environment and maintains a national dose register.
- Waste: collects and reports national statistics on waste generation and management, including Ireland's progress towards EU waste targets.

The EPA's monitoring and assessment programmes help fulfil several statutory reporting duties to the EU and national government. The EPA also makes data accessible to the public through the national open data portal (data.gov.ie), its reports such as the State of the Environment report, annual reports on drinking water, urban waste water, bathing water, water quality, air quality, greenhouse gas emission inventories and projections and waste statistics, and its websites such as catchments.ie and beaches.ie.

The EPA, unlike many of its sister organisations in Europe, has statutory responsibility for co-ordinating a national research programme in the area of environmental protection. This function involves an annual call for proposals from universities and other institutions for research in areas identified by the EPA and others (including government departments) as being of national priority. In 2018, EPA invested almost EUR 10 million in research.

The overarching objective of the research programme is to use knowledge to protect and improve the natural environment and human health. The research strategy for 2014-2020 is built around three pillars: climate, water and sustainability (which includes radiological protection). In each of these pillars, the focus of the research is to:

- Identify pressures: by providing assessments of current environmental status and future trends to identify pressures on the environment;
- Inform policy: by generating evidence, reviewing practices and building models to inform policy development and implementation; and
- Develop solutions: by using novel technologies and methods that address environmental challenges and provide green economic opportunities.

The research programme has three National Research Co-ordination Groups, one for each pillar of research, that convene national stakeholders including government departments, agencies and NGOs. The EPA also invites government department staff and experts from other bodies to participate in research steering groups. Information on EPA-funded research is disseminated via several channels, including DROPLET (a database on water research), social media platforms such as Twitter (> 5 000 followers) and LinkedIn, research platforms such as ResearchGate, as well as newsletters, events and webinars. The

EPA also co-funds research with other funding agencies and links up Irish researchers to relevant Horizon 2020 funding, EU Joint Programme Initiative funding and other EU funding streams.

Advocacy functions

The EPA strategic plan sets out the goal to be an effective advocate and partner. The EPA aims to "work with others to advocate for a clean, productive and well protected environment and for sustainable environmental behaviour". Advocacy functions are split between several offices but some directors lead on behalf of the Board with organisations that interface with many offices within EPA (such as Irish Water, Bord Bia (the Irish Food Board)). The EPA does not have an overarching advocacy strategy but has guidelines for staff on advocacy and partnering.

Examples of this function include:

- A number of advocacy campaigns under the framework of the National Waste Prevention Programme. The EPA delivers the programme through partnerships with local authorities, regional waste offices, government agencies and public bodies as well as sectoral groups and bodies to promote the circular economy and enable waste prevention actions by business, public sector and communities. Campaigns to improve resource efficiency have included Smart Farming, in partnership with the Irish Farmers' Association, and StopFoodWaste.ie.
- The EPA has a key role in the development and implementation of the National Radon Control Strategy and through this has undertaken a range of advocacy campaigns and activities to minimise the exposure of members of the public to radon gas in their homes and workplaces. The EPA has also created a dedicated website (http://www.radon.ie/) which provides customised information for different groups (such as homeowners, medical professionals and local authorities). This website was launched in 2016 and received almost 100 000 views in both 2017 and 2018.
- Publicising data and reports, such as the State of the Environment Report, through media interviews, press releases and public events.
- Dedicated websites, such as http://www.catchments.ie/ that shares science and stories about Ireland's water catchments and people's connections to their water.
- Conferences, public lectures (e.g. Climate Lecture series) and workshops.
- "Citizen science" initiatives to increase awareness and involvement of the public in the areas of clean air, clean water and sustainability.
- Competitions targeting youth, such as "The story of your stuff" (http://www.thestoryofyourstuff.ie/).
- Co-ordinating the National Dialogue on Climate Action.

The EPA has a corporate communications strategy 2017-2020 that identifies target audiences, desired behaviour changes, corporate messages, risks, and communications channels. In 2015 the EPA commissioned an external communications audit to gather information about perceptions of the EPA and the effectiveness of its communications with target audiences. The audit found that the EPA is viewed as a high-calibre and trusted source of information but recommended that the EPA address the issue of 'infobesity' in its communications (use of acronyms and technical terminology) and the lack of clear champions and clear voices on EPA issues. Additional polling commissioned by the EPA revealed that awareness of the EPA and its work was relatively lower among younger audiences and so are exploring way to reach this target group.

In addition to its regulatory, knowledge and advocacy functions, the EPA provides the secretariat to the Climate Change Advisory Council as well as technical and administrative support to the National Dialogue on Climate Action.

Institutional co-ordination

The EPA liaises with many government bodies, regulatory authorities and other stakeholders (Table 2.1). Formal arrangements, in the form of Memoranda of Understanding (MoU), are in place with 23 organisations (in 21 MoUs) from Ireland, the United Kingdom and the European Union (Box 2.2). The Board receives an update on all MoUs annually. MoUs are published on the EPA website where agreed with the other party.

Several co-ordination mechanisms that are not statutory requirements or defined under formal arrangements are also in place. For example, the EPA's water team meet with the Commission for Regulation of Utilities (CRU) on a quarterly basis to exchange and update each other on their respective regulatory roles in relation to Irish Water. Personal relationships appear to be an important factor in the degree of co-ordination between the EPA and other institutions, supplementing formal arrangements.

> Box 2.2. Memoranda of Understanding
>
> The EPA has MoUs in place with the following organisations:
> - An Bord Pleanála
> - ASN (*Autorité de sûreté nucléaire*) – French Nuclear Safety Authority
> - Bord Gais Networks
> - Commission for Regulation of Utilities
> - Central Statistics Office
> - Climate Change Advisory Council
> - Department of Agriculture, Food & the Marine
> - Department of Communications, Climate Action & Environment, Kilkenny County Council and Galmoy Mines
> - European Atomic Energy Community
> - Food Safety Authority of Ireland
> - Health & Safety Authority
> - Health Service Executive
> - Marine Institute
> - Met Eireann
> - National Directorate for Fire and Emergency Management
> - National Parks & Wildlife Service of the Department of Arts, Heritage and the Gaeltacht
> - Sustainable Energy Authority of Ireland
> - UK Drinking Water Regulators
> - Office of Public Works
> - Irish Coast Guard
> - Office for Nuclear Regulation (United Kingdom)
>
> Source: Information supplied by EPA.

The EPA co-ordinates with Ireland's 31 local authorities, which have significant environmental protection responsibilities. Since 2004, the EPA and local authorities have operated an administrative network – the Network for Ireland's Environmental Compliance and Enforcement (NIECE) – which provides the framework for both oversight and support.

Several interviewees referred to an inherent tension in the relationship between the EPA and local authorities. Upon its establishment, the EPA assumed many of the responsibilities for environmental protection previously held by local authorities and over time has taken over additional functions. The dual role of regulator and provider of advice and assistance can create tensions as the EPA moves between the different roles.

Legislation also empowers the EPA to co-ordinate with relevant European institutions. Section 52 of the EPA Act 1992 lists liaison with the European Environment Agency among the primary functions of the EPA, provided for under Council Regulation 1210/90/EEC.

Table 2.1. Co-ordination with other state and regulatory bodies

Authority	Area/sector of co-ordination	Formal co-ordination arrangement in place?
Department of Communications, Climate Action and Environment (DCCAE)	Climate action & greenhouse gas emissions Circular economy & waste prevention Chemicals and market surveillance Ozone depleting substances (ODS) and F Gas Environmental licensing including GMOs and Dumping at Sea Noise (environmental noise policy) Radiation Waste Waste statistics Research	Oversight Agreement and associated Performance Delivery Agreement The Network for Ireland's Environmental Compliance and Enforcement (NIECE) Research covered by Memorandum of Funding for Research Programme Research Co-ordination Group
The Department of Housing, Planning and Local Government (DHPLG)	Drinking water Urban waste water Water quality Microbeads Water quantity Water Abstractions Noise (local authority-controlled roads; planning and development policy, guidance and standards Strategic Environmental Assessments Emergency Planning EIA legislation/environmental licensing Research	Covered under DCCAE Oversight Agreement and associated Performance Delivery Agreement Water Framework Directive established co-ordination structures: Water Policy Advisory Committee (WPAC); National Co-ordination & Management Committee (NCMC); National Technical Implementation Group (NTIG); Memorandum of Funding for WFD related work NIECE SEA Governance Forum Research Co-ordination Group
Department of Business, Enterprise and Innovation (DBEI)	Radiation (carriage of radioactive materials by road) Chemicals/Market Surveillance Research	Research co-ordination through national committees for Innovation 2020 and Horizon 2020 Research Co-ordination Group
Department of Transport, Tourism and Sport	Noise (airport, rail, motorway and primary road network) EPA comments on SEAs Research	Research Co-ordination Group
Central Statistics Office (CSO)	Climate Action (Greenhouse gas emissions)	MoU Research Co-ordination Group

Authority	Area/sector of co-ordination	Formal co-ordination arrangement in place?
Department of Agriculture, Food and the Marine (DAFM)	Waste/ Circular Economy Research Climate action & greenhouse gas emissions Intensive agriculture Chemicals Waste Water quality EPA comments on SEAs Research	MoU WPAC NTIG National Pesticides in Drinking Water Action Group (NPDWAG) NIECE Research Co-ordination Group
Local authorities	EPA has supervisory role Circular economy and waste prevention Air quality (smoky coal, and air quality monitoring) Chemicals Drinking water Industrial emissions Waste Wastewater Water quality (catchment protection) Research	NIECE WFD governance structures (WPAC, NCMC, NTIG and five regional committees) Research Co-ordination Group
An Bord Pleanála (ABP)	Environmental Impact Assessments Consultations on environmental licensing Projects of Common Interest – large scale energy infrastructure projects with cross border aspects	MoU
Commission for Regulation of Utilities (CRU)	Climate action Energy safety Drinking water	MoU
Health Service Executive (HSE)	Drinking water Bathing water Health and environment Environmental licensing Research	MoU NIECE Health Advisory Committee Research Co-ordination Group
Health and Safety Authority (HSA)	Chemicals Radiation (radon in workplaces; carriage of radioactive materials by road) Seveso Directive/Industrial licensing	MoU Research Co-ordination Group
Health Information and Quality Authority (HIQA)	Radiation (regulation of medical exposures)	MoU in prep
All other plan-making government departments	EPA comments on SEAs	No
An Garda Síochána	Radiation (security of radioactive sources) Waste (waste crime) Emergency planning & response (radiation)	Formal work plan and policy statement National Waste Enforcement Steering Committee NIECE
Inland fisheries Ireland	Water quality (where pollution incidents impact fish) Environmental licensing	WFD governance structures: NIECE, WPAC, NTIG and five regional committees
Marine Institute	Water quality Chemicals Research	MoU WFD Governance Structures Research Co-ordination Group
Office of Public Works	Water quality Research	MoU Research Co-ordination Group
Revenue (Customs)	Chemicals	NIECE

Authority	Area/sector of co-ordination	Formal co-ordination arrangement in place?
Regional Waste Management Planning Offices	Circular economy & waste prevention Waste	NIECE
Waste Enforcement Regional Lead Authorities (WERLA)	Waste	NIECE National Waste Enforcement Steering Committee
National Transfrontier Shipment of Waste Office (NTFSO)	Waste	NIECE
National Waste Collection Permit Office (NWCPO)	Waste	NIECE
Food Safety Authority of Ireland (FSAI)	Emergency planning & response (radiation) Radiation monitoring Chemicals Research	MoU Health Advisory Committee Research Co-ordination Group NIECE
Met Eireann	Emergency planning & response (radiation) Radiation monitoring Air quality, monitoring, modelling & forecasting	MoU Research Co-ordination Group
Teagasc	Climate action & greenhouse gas emissions Research	Data provider named in Ireland's National Inventory System, Research Co-ordination Group, Steering committee for the Agricultural Catchment Programme
Sustainable Energy Authority of Ireland	Climate action & greenhouse gas emissions Research	Data provider named in Ireland's National Inventory System, Research Co-ordination Group
Department of Culture, Heritage and the Gaeltacht - Built Heritage and Architectural Policy Unit, Irish Research Council, Irish Water, National Parks and Wildlife Services, National Transport Authority, Science Foundation Ireland, Enterprise Ireland, Geological Survey of Ireland, National Economic and Social Council	Research	Research Co-ordination Group

Source: Information provided by EPA.

Relations with executive

The Department of Communications, Climate Action and Environment (DCCAE) has been the EPA's sponsor in government since 2016. Prior to this, the EPA had been under the aegis of the Department of the Environment, Community and Local Government (now re-named the Department for Housing, Planning and Local Government, DHPLG). DHPLG retains responsibility for water policy, however, and so the EPA has an important relationship with DHPLG in this policy area.

A tripartite Oversight Agreement and associated Performance Delivery Agreement is in place between DCCAE, DHPLG and the EPA. This Oversight Agreement recognises that DHPLG has responsibility for a number of areas of direct relevance to the EPA's remit (for water and planning related issues). The Oversight Agreement addresses the following:

- Legal framework
- Operational environment of the EPA
- Purpose and responsibilities of the EPA
- Compliance with the *Code of Practice for the Governance of State Bodies 2016* (DPER, 2016[4])
- Arrangements for oversight, reporting and monitoring
- Mutual commitments
- Performance Delivery Agreement
- Duration of agreement

The EPA also engages with other government departments including the Department of Agriculture, Food and Marine; the Department of Health; the Department of Business, Enterprise and Innovation; the Department of Culture, Heritage and the Gaeltacht; the Department of Transport, Tourism and Sport; and the Department of Public Expenditure and Reform.

Input into policy development

The EPA does not set government policy but aims to support and influence the policy development process through its scientific data, evidence and knowledge. The EPA Act 1992 empowers the EPA to advise "of its own volition" the government on environmental protection and related matters. This function encompasses giving advice to the government on any proposals for legislative change or other policy matters, as well as reporting and making recommendations on particular environmental issues or problems.[1] The EPA's strategic action plan 2016-2020 sets out specific actions related to policy development, notably: co-ordinating EPA scientific expertise to support the development of new policy (e.g. clean air); ensuring that new legislation is implementable (e.g. the transposition of EURATOM BSS); and using the opportunities to build competence in engaging on legislative development. The EPA gives particular emphasis to its role in providing the evidence base to inform policy, drawing on its well-respected data and reports. Results of EPA-funded research also contribute to the body of evidence that can inform policy.

The EPA informs the policy-making process through a number of channels. The EPA makes formal submissions on draft legislation put out for public consultation; presents to committees of parliament on a wide range of topics that then informs debate and discussion; and responds to requests, often informal, from government departments on the development of policy and draft legislation, particularly where technical or scientific input is required. Comments on policies or plans that are released for public consultation are published and available to the public. Where the EPA comments on EU matters, such as drafts of directives, the comments are considered by the departments and incorporated into the official Irish response. Informal consultations are not made public.

The EPA is also involved in a number of European and other international structures which feed policy development, providing technical and scientific advice while department officials address policy issues. These include EU policy fora to develop the *acquis* (in support of various departments and permanent representations); EU scientific working groups that inform the implementation of EU legislation and standards; and EU committees in which the EPA acts as technical expert to support department officials who are responsible for formal voting. The EPA also represents Ireland at international negotiations in support of the government, including the UNFCCC COPs, the IPCC and the Kyoto, Stockholm, Basel, Rotterdam and Montreal Conventions.

There is a more direct link into supporting the policy-making process on nuclear and radiological issues. DCCAE relies primarily on the EPA for input into this policy area as there is no dedicated technical team within the parent department. The EPA advises on policy and provides technical input and support for the negotiation of international agreements and conventions sponsored by the International Atomic Energy Agency (IAEA) and Euratom.

The amount of time and resources dedicated to policy work varies across offices and over time. At times a significant workload is associated with revising legislation. For example, 20% of staff time in the Office for Radiation Protection and Environmental Monitoring was dedicated to supporting the transposition EU legislation in 2017 and 2018. This contrasts with input to the policy process for air quality, for example, which requires an estimated three person days per year. To a large extent the resource requirements reflect the policy priorities of the government department at any particular time.

The EPA can request changes in legislation where implementation challenges are faced or market failures identified. The most recent instance was in May 2016 when, at the request of DCCAE, the EPA provided proposals for legislative change in several areas. On that occasion, some minor suggestions related to administrative anomalies or unclear drafting were incorporated into the draft legislation, but more significant suggestions were not. There is limited evidence that the EPA has influenced changes in legislation where there is not an overriding policy imperative e.g. fees for licence reviews or technical amendments. However, there is evidence where there are requests from government departments to the EPA when there are new policy or amendments to legislation.

Strategic objectives and planning

EPA operates in the framework of a strategic plan 2016-2020, *Our Environment, Our Wellbeing*, which sets out the EPA's goals to be:

- a trusted environmental regulator
- a leader in environmental evidence and knowledge
- an effective advocate and partner
- responding to key environmental challenges
- organisationally excellent

It is the EPA's fifth strategic plan and was published when Ireland was seeing the signs of an economic recovery. It is designed to deliver on the EPA's vision of "a clean, healthy and well protected environment supporting a sustainable society and economy" and its mission "to protect and improve the environment as a valuable asset for the people of Ireland; to protect our people and the environment from the harmful effects of radiation and pollution" (Figure 2.2). Under each goal, the plan identifies the main objectives, expected outcomes, the actions that the EPA will take to achieve these outcomes and an owner responsible for each action (Table 2.2).

The Director General, in consultation with the senior management team and staff of the EPA, sets the objectives. Input from the EPA advisory committee is also considered. A draft of the 2016-2020 strategic plan was published on the EPA website for public consultation. The EPA published a summary on its website (http://www.epa.ie/pubs/reports/other/corporate/occs/Consultation_Issues_Response.pdf) of the major issues that were raised through the consultation and an explanation as to how the EPA can respond to them.

The 2016-2020 strategic plan was submitted to the former Minister for Environment, Community & Local Government. The EPA corporate governance manual states that "a copy of the draft strategic plan should be sent for views from the Minister or Department who should have up to 12 weeks to comment".

Figure 2.2. Linking EPA's vision and mission with its annual work programme

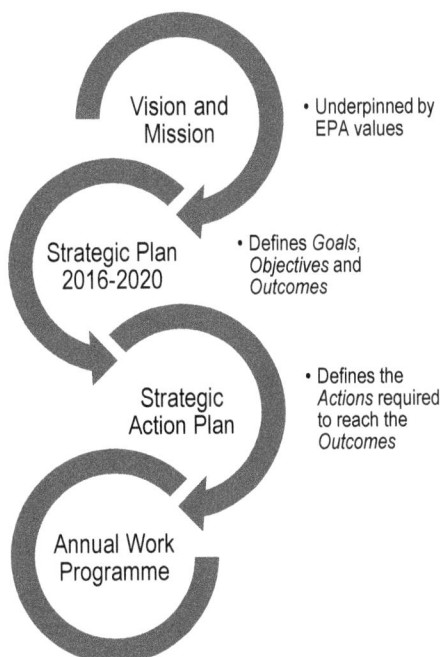

The action plan is translated into annual work programmes that are also informed by a number of other documents. In the last quarter of each year, each office prepares a work programme identifying a number of defined work areas and outputs. These work programmes are informed primarily by statutory functions, new legislation and commitments arising from the Public Sector Reform Plan. In addition to the strategic plan, the following organisational strategies are considered in the preparation of the EPA's annual work programme:

- Oversight Agreement & Performance Delivery Agreement
- Corporate and Office Risk Registers
- ICT Strategy
- Communications Strategy
- Human Resources & Development Strategy
- Annual Communication Plan.
- Annual Internal Audit Plans
- Internal Audit Implementation Plans
- Annual Budget

The annual work programme focuses on outputs linked to the strategic goals and objectives, target dates and ownership and is submitted to the Board for approval. Monthly progress reports on the work programme KPIs are submitted to the Board by the relevant director.

In 2018 the EPA carried out a mid-term review of the 2016-2020 strategic plan through a collaborative process involving over 40 members of staff. The review resulted in amendments to the objectives – to include references to the circular economy and waste management – and to the outcomes – to include "a climate resilient society and economy" and "strengthened national waste enforcement structures". A number of actions were also revised.

Table 2.2. EPA's strategic plan 2016-2020

Goal	Objectives	Outcomes by 2020
Trusted Environmental Regulator	Ensure the on-going development of a proportionate and effective regulatory approach. Align EPA resources to target interventions and reduce environmental risk	A risk-based, responsive regulatory approach that engages stakeholders and protects the environment and people. Reduced environmental risks at EPA regulated facilities through tailored interventions and by ensuring appropriate financial provisions are in place. Driven the improved delivery and management of water and waste infrastructure.
Leader in Environmental Evidence & Knowledge	Realise the full potential of the EPA's knowledge, skill, expertise and regional presence as key national resources in the protection of the environment and human health. Accelerate the provision of timely and tailored information to meet the specific needs of stakeholder groups.	More timely evidence-based environmental assessments to inform policy and decision making at national, regional and local levels. Better provision of online, up-to-date and accessible information on the environment to stakeholders. A research programme that addresses knowledge gaps and helps identify solutions to emerging and complex environmental problems.
Effective Advocate and Partner	Strengthen the EPA's capability and capacity to influence, advocate and partner to help achieve a clean, healthy and well protected environment. Engage the public in the protection and improvement of the environment. Promote a greater awareness of the impact of environment quality on human health.	Targeted opportunities to integrate environmental priorities and sustainability into sectoral, economic and social policies. Developed public participation programmes to increase awareness of environmental issues and support the engagement of the public in environmental protection. Collaborated with health agencies and other bodies to realise the benefits of a good environment for health and wellbeing.
Responding to Key Environmental Challenges	Tackle the challenges to deliver improved water quality in Ireland. Engage with other strategic partners to promote the development of a holistic national response to climate change. Enhance air and radiation protection in Ireland.	Effective and resilient structures in place to deliver better outcomes for water quality. Established a climate change secretariat in the EPA as a centre of excellence that supports the national transition to a low carbon economy. Strengthened the air quality and radiation protection frameworks to further protect people and the environment.
Organisationally excellent	Develop our staff and align our organisation to deliver best environmental outcomes. Focus on the development and promotion of organisational health, wellbeing and safety at work. Promote a culture of leadership, reform and innovation.	Our functions and resources aligned to be responsive and adaptable to meet emerging challenges. Engaged all staff to foster a supportive workplace environment. Enhanced capacity in the area of organisational change and in the use of ICT to support reform and innovation.

Source: (EPA, 2018[5]), EPA Strategic Plan 2016-2020 Our Environment, Our Wellbeing. http://www.epa.ie/pubs/reports/other/corporate/EPA_StrategicPlanWeb_2018.pdf.

Independence

The Oversight Agreement between DCCAE, DHPLG and the EPA states that the EPA "operates as a non-commercial, regulatory body. It has independence in the performance of certain functions as set out in the 1992 Act, and other relevant legislation".

The EPA Board makes operational decisions, such as licensing decisions, without the involvement of government departments or other bodies. Furthermore, Section 40 of the EPA Act states that a person may not communicate with any individual within the EPA, its Advisory Committee or any committee established by the EPA, for the purpose of influencing improperly their consideration of any matter that falls to be considered or decided by the EPA, committee or consultative group. There has never been a formal complaint to the EPA in this regard.

Similarly, strategic decisions do not require approval by the legislature or the executive although EPA's strategic plan is submitted to DCCAE for comment. The updated *Code of Practice for the Governance of State Bodies* (DPER, 2016[4]) published in 2016 stipulates that the strategy of non-commercial state bodies should align with the specific objectives in the parent department's strategy (to the extent relevant). It allows the minister twelve weeks to revert with views on the EPA's draft strategy.

The EPA can make recruitment decisions within its headcount, publish reports and set enforcement charges for regulated entities independently.

The EPA is largely dependent on the government for budget and staff (Box 2.3). The EPA negotiates its budget with its parent departments on an annual basis. In 2018, around 79% of the EPA's income came from government sources (DCCAE, DHPLG and the Environment Fund). The EPA also requires the agreement of DCCAE and subsequently the Department of Public Expenditure and Reform (DPER) to make any changes to staff numbers and grades.

The Government appoints the EPA Director General after selection by a committee whose members are defined in statute (Section 21 of the EPA Act, 1992), namely: the Secretary to the Government; the Secretary of DCCAE; the Chair of the National Trust of Ireland; the Managing Director of the Industrial Development Authority; the General Secretary of the Irish Congress of Trade Unions; and the Chief Executive of the Council for the Status of Women. The function of the Director General is to ensure the efficient discharge of the business of the agency and to arrange the distribution of the business of the agency among its directors (Section 23, EPA Act).

The government also appoints directors of the EPA (Section 24), who form the Executive Board alongside the Director General. A director of the EPA is a full time position, based at the EPA's headquarters in Wexford, with a five-year term of office. Appointment follows from a recommendation made by an independent selection committee convened in accordance with the EPA Act, 1992. The independent selection committee conducts a public recruitment competition to select candidates suitable for recommendations to Government. A director may be reappointed by government for a second or subsequent term of office for five years or less. Directors report to the Director General, as specified in the advertisements for director posts.

According to statute, the government may remove the Director General or directors from office "if, in their opinion, [they] have become incapable through ill-health of effectively performing [their] duties, or for stated misbehaviour, or if [their] removal appears to the Government to be necessary or desirable for the effective performance by the Agency of its functions" (Sections 21:16 and 24:12). In such cases, the government must provide a written statement of the reasons for removal to each House of the Oireachtas.

> **Box 2.3. Creating a culture of independence**
>
> (OECD, 2017[6]) explores how to establish and implement independence with regulators. Independence comes in two forms: de jure independence refers to the formal independence granted by law, whereas de facto independence promotes practical independence as shown by actions, decisions and behaviours.
>
> **Figure 2.3. The five dimensions of independence identified by the Guidance**
>
>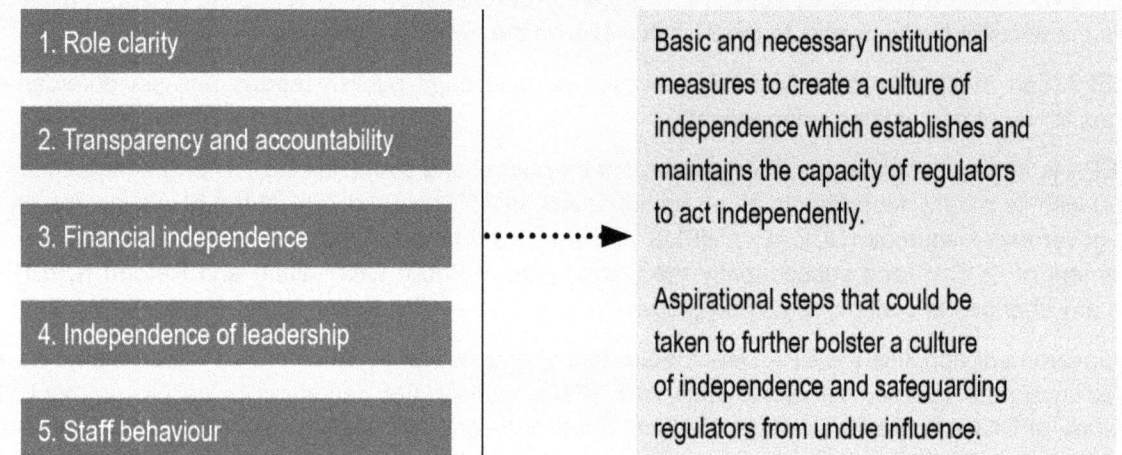
>
> Each of the five dimensions includes practical guidelines that can be considered as the basic and necessary institutional measures to create a culture of independence which establishes and maintains the capacity of regulators to act independently, based on an analysis of regulators' institutional processes and practices within the OECD Network of Economic Regulators (NER). The guidelines also include a set of aspirational steps that could be taken to bolster a culture of independence and safeguarding regulators from undue influence.
>
> Source: (OECD, 2017[6]), Creating a Culture of Independence: Practical Guidance against Undue Influence, The Governance of Regulators, Paris, https://doi.org/10.1787/9789264274198-en.

Input

Financial resources

The EPA obtains the majority of its funding from government sources. The majority of government funds are appropriated through DCCAE, while DHPLG has provided between 10.5% and 13% of the yearly budget over the last three years to deliver on priorities related to the Water Framework Directive (WFD).

The budget is allocated by major spending category (i.e. pay, pension, and research) or for specific deliverables (i.e. WFD delivery, emergency response, research, and the climate secretariat and climate dialogue). The budget takes into account funds necessary to finance agreed upon programmes of work, but the budget lines are not presented according to the priorities listed in the strategic objectives.

EPA receives funding from four sources:

- Exchequer income: funds provided by DCCAE and DHPLG. These funds require yearly support from DCCAE and are approved by the DPER. Funds from DHPLG can only be used for water-related activities – including the operation of regional laboratories for monitoring under the WFD – according to a Memorandum of Funding, and not for staff costs without sanction. Research is also partly funded by exchequer income.
- Environment Fund: levies collected by the national government from the plastic bag levy and landfill levy. The allocation of this fund is determined and approved by DCCAE. In each of the last three years in the estimates letters, the EPA has signalled its concern about the sustainability of using this fund for non-pay, non-discretionary expenditures (i.e. operational costs related to light, heat, rent, insurance, etc.). In 2018, EUR 4.6 million of the fund was allocated to these non-discretionary expenditures, falling to EUR 1.4 million in 2019. This has been accompanied with requests each year to move these expenditures off the fund. Research is also partly funded by the Environment Fund.
- Earned income: levies from licensing fees, radiological income and enforcement (prosecution) income. Licensing fees are set in legislation, requiring an amendment by the Oireachtas to change, and generally cover 10-15% of the cost of licensing. Losses are covered by Exchequer funding. Enforcement income is calculated in the first instance via a cost model that costs each section of activities, including staff time and overhead costs. Each team then calculate the cost associated with their activities that are chargeable. These charges are then applied to the activities and charged out accordingly. Most of the enforcement activities achieve full cost recovery, while the EPA is making efforts to move those that do not towards full cost recovery. Each year a memo is prepared by each of the teams responsible for charging to outline how they propose to charge for their activities in the coming year. Maximum fines from prosecution are set by the Oireachtas in the governing legislation (see Enforcement section). The fines imposed in each case are specific to the case and at the discretion of the judge. Execution of earned income is approved by the EPA Board.
- Other income: earned from emission trading units costs recovery, staff pension contributions and other/sundry income. Execution of other income is approved by the EPA Board.

Table 2.3. EPA Budget by category, real (EUR millions) and percentage of total income

	2015	2016	2017	2018
Total income	59.0	59.8	63.0	65.0
Exchequer income	26.9 (45.7%)	33.5 (56.1%)	39.9 (63.4%)	42.3 (65%)
Environmental Fund income	16.1 (27.3%)	12.8 (21.4%)	9.8 (15.5%)	9.0 (13.8%)
Earned income	13.3 (22.5%)	10.7 (18.0%)	10.8 (17.2%)	10.9 (16.7%)
Other income	2.6 (4.5%)	2.7 (4.6%)	2.5 (3.9%)	2.9 (4.5%)
Total STATE Income	43.1 (73.0%)	46.3 (77.5%)	49.7 (78.8%)	51.3 (78.8%)
Total EARNED & OTHER Income	15.9 (27.0%)	13.5 (22.5%)	13.3 (21.2%)	13.8 (21.2%)

Source: Information provided by the EPA, 2019.

Managing financial resources

Budgets are decided on a yearly basis. The EPA provides budget estimates to its parent departments for that year. In 2018, the EPA was asked to provide a high-level three-year estimate for the first time, but has not been requested to do this again. The EPA operates a "business partnering approach", whereby EPA managers responsible for each programme area are tasked with negotiating their portfolio of funds for the following year, supported by the Finance and Organisational Services Programme within OCCS. The

Performance Delivery Agreement, a high-level tripartite agreement between the EPA, DCCAE and DHPLG, provides an opportunity for the three bodies to come together and discuss common challenges, including funding.

For Exchequer funds, the EPA provides an estimate of its budget requirements for the following year in July/August to DCCAE and DHPLG via an estimates letter. After reviewing the request, which includes an option to potentially adjust the request, DCCAE and DHPLG include the EPA requirements in their overall estimates requested to DPER. DPER review and present to the Dáil for approval in October, with formal notification being received in December/January for the one-year estimate. The three-year estimate in 2018 did not receive any signal or commitment from government. EPA funding requirements from the Environmental Fund are included in the estimates letter submitted to DCCAE and are subject to approval by the Ministry. Ring fenced budgets for specific areas such as the Office of Environmental Enforcement, research, the National Waste Prevention Programme and the Climate Dialogue are also assigned. These are all governed by memoranda of funding agreed with the relevant departments.

The estimates process is underpinned by the regular interaction and planning between the different offices and programme areas in the EPA and the relevant government department, on programme activities such as research, the National Waste Prevention Programme and the WFD. This planning information is referenced by DCCAE and DHPLG when reviewing the EPA estimates letter and is key to obtaining approval.

The EPA requires approval for budget allocation and spending on items funded from either the Exchequer or the Environment Fund. Approval is required from the relevant government department prior to drawing down funds to cover actual expenditure from these two budgetary sources. Funds from the Exchequer or Environmental Fund cannot be carried forward. The EPA can and does carry forward earned income, amounting to on average 1% yearly over the three years. The EPA communicates the predicted carry forward figures to DCCAE and publishes the overall actual year end carry forward position in the EPA annual report.

The Budget is communicated to all budget holders and electronic reports on expenditure versus budget are available to all budget holders on the Integra system. Expenditure against Budget is reported to the EPA Board on a monthly basis via the Financial Management report. The internal budget is subject to two internal revisions in June and September, which are subject to EPA Board approval. DCCAE are provided monthly updates on actual expenditures.

Section 50 of the EPA Act requires the EPA to keep proper accounts of all funds it receives or expends. The EPA Board is responsible for preparing financial statements. As part of this duty, the Annual Report must contain an Annual Financial Statement. This Statement is audited by the Comptroller and Auditor General (C&AG) of Ireland, who then reports the audit to the Public Accounts Committee (PAC). The EPA can be called to appear before the PAC to report on how it managed the resources at its disposal and on any other matter of interest to the Committee. The EPA is rarely called to attend the Public Accounts Committee (PAC) to defend its annual reports, having only occurred three times since it was established. When it was last called in April 2019, the chair of the PAC noted the EPA's excellence in preparing its annual reports. The EPA C&AG audited accounts are also presented to the Minister, who then reports on these accounts to the Oireachtas. Under the *Code of Practice for the Governance of State Bodies* (DPER, 2016[4]), the EPA has a duty to provide draft unaudited annual accounts to DCCAE and DPER within two months of the end of the financial year.

In accordance with Section 7.3 of the *Code of Practice for the Governance of State Bodies*, the EPA Board is responsible for ensuring that effective systems of internal control are instituted and implemented, including for financial management.

For procurement, the EPA is required to adhere to the *Public Procurement Guidelines for Goods and Services*, published by the Office of Government Procurement in 2017. It is the responsibility of the Board to ensure these procedures are adhered and fully conversant with the current value thresholds for the application of EU and national procurement rules.[2] In 2008, a Procurement Officer role was established, who developed guidelines and templates for budget holders who are involved in procurement and publishes a quarterly Procurement Bulletin to all staff.

Human resources

The EPA approved staff complement is currently 420 staff members. As of September 2019, 416 posts were filled, of which 262 (63%) are technical staff (i.e. engineers, scientists, specialised researchers) and 154 (37%) are considered management or support (Table 2.4).

Table 2.4. Staff by category, 2018

Staff category	Female	Male	Total
Management	31	36	67
Technical staff	139	123	262
Support staff	76	11	87
Total	246	170	416

Notes: Management includes all managers (senior, technical and administrative) at a Level 2/3 and upwards. Total reflects staff as of September 2019, which totals 416. The remaining posts are currently in competition to be filled.
Source: Information provided by the EPA, 2019.

The 416 posts are filled by permanent employees of the EPA. In addition, the EPA hires contractors to provide certain services or support project areas where expertise is not developed or required in-house. Contractors are used in different parts of the EPA and to different extents. The list below includes areas of significant contractor use, but contractors can be used from large projects to small support projects (< EUR 5 000). Areas of high contractor use include:

- ICT programme: 21 staff and 38 contractors. The most heavily dependent areas on contractors and external consultants are Development (1:8.6 staff to contractor ratio), Service Desk (1:7.8 staff to contractor ratio) and Data management (1:2.2 staff to contractor ratio).
- Chemicals work stream within the OES programme: four staff members and up to a maximum of five external contractors/consultants engaged at project intervals per year, in accordance with Framework Agreements.
- Pollutant Release and Transfer Register (PRTR) work stream within the OES programme: one staff member (4.5 days/week) and one full-time contractor
- Legal services: Fully procured via third party law firms, equivalent to five full time employees. Outside counsel is used to ensure skills and knowledge of environmental law are kept up-to-date.

The EPA has developed a Human Resources Development (HRD) Strategic Framework and Action Plan 2017-2021 (Table 2.5). The aim is for EPA to "become a role model for the stewardship and development of our people and organizational resources". The overarching theme for the HRD strategy is "Engaging, Enabling, Empowering" and is supported by four strategic goals:

- Foster a healthy, engaged, and resilient workforce
- Develop our people and organisational resources
- Empower our managers as experts and leaders
- Evolve our HR delivery model

Each of the four goals is supported by four to five strategic priorities, as well as a high level outcome that the EPA intends to achieve by 2021. The KPIs are the actions noted in the plan. Each strategic priority is supported by four to six actions in the action plan, which are prioritised for the 2017-19 period of the strategy.

The HRD strategy was developed through extensive collaboration with staff from across the EPA. The first cycle focused on the development of an initial strategic framework through a series of workshops with the Board, the Human Resources Learning and Development team and the Senior Management Network (SMN). The second cycle focused on refining the strategic framework and developing a supporting action plan through a series of four one-day workshops focused on the emerging strategic themes followed by a full peer review by the SMN and project team members. Wider staff consultation and engagement was maintained throughout the process directly with staff as well as a series of presentations delivered throughout the roadshows.

Table 2.5. EPA Human Resources Development Strategic Framework 2017-2021

Goal	Strategic priorities	Outcome
1. Foster a healthy, engaged, and resilient workforce	1. Foster a workplace in which the health and wellbeing of our people remains a key priority. 2. Nurture a highly-engaged and resilient culture that promotes effective collaboration across organisational boundaries. 3. Strengthen the change management capabilities of individuals and teams. 4. Promote and embed a spirit of innovation throughout the workplace.	By 2021, we will have a healthy, engaged and resilient workforce that thrives on innovation and change.
2. Develop our people and organisational resources	1. Implement an approach to strategic resource management that incorporates staff and contractors. 2. Develop an updated competency framework to support every stage of the employee life-cycle. 3. Amend our approach to recruitment to take account of changing competencies, ensure equity and fairness, and address changing needs in current recruitment processes. 4. Introduce a new approach to career planning and development for all staff. 5. Use a broader range of approaches to people development that are better aligned with the updated competency framework and the needs of individuals and teams.	By 2021, we will have a workplace that encourages individuals and teams to excel in realising EPA's strategic goals and objectives.
3. Empower our managers as experts and leaders	1. Develop and implement the expert as leader framework in support of people management and development. 2. Promote the practice of developmental conversations between managers and their staff. 3. Realign internal systems, procedures, and practices to support the new focus on people management and development. 4. Refine the EPA's system of incentives to support the expert as leader framework.	By 2021, we will have empowered our managers as technical experts and leaders in people management and development.
4. Evolve our HR delivery model	1. Strengthen strategic HR leadership across the EPA. 2. Build a unified HR team that is a role model for excellence in HR practice across the EPA. 3. Execute a root and branch review of our HR process model with a view to simplifying and streamlining our work system. 4. Develop a network of middle managers as a principal means to strengthen the adoption of HR practices. 5. Cultivate a strong responsive culture to support the adoption of modern HR practices.	By 2021, we will have a HR delivery model that promotes sound stewardship and development of our people and organisational resources.

According to Section 29 of the EPA Act, the EPA is empowered to appoint staff subject to the numbers and grades sanctioned by DCCAE with the consent of DPER. Therefore, increases in the headcount require approval from DCCAE and DPER. The EPA usually advocates for an increase in headcount when it takes on new functions or there is a growth in existing functions, but not outside of these occasions.

DCCAE, in turn, makes the case for a general head count for all government agencies under its purview. DCCAE engages with DPER with regards to EPA staff numbers.

The EPA submits an annual workforce plan to the Assistant Secretary General of DCCAE, which is produced by HR in conjunction with the directors of the EPA and makes the business case for staffing. In preparing the most recent workforce plan, each office in the EPA was tasked with a review of existing resources and functions to ensure that staff were used to best effect across the organisation. This included a review of skills and competencies required to maintain EPA's standard of service to the parent departments and stakeholders. The EPA also looked at a range of options including outsourcing of certain business services, streamlining of existing processes and use of integrated digital technology platforms and partnering. The review highlighted a number of skills gaps that are emerging in light of new and emerging legislative functions. Business cases were prepared to advocate for new posts in the EPA, with the majority to provide more resources for processing licensing applications and for water-related functions.

Table 2.6. Workforce movement at the EPA, 2016-18

Category	2016	2017	2018
Recruitments	44	19	47
Promotions	25	21	31
Resignations	10	9	14

Source: Information provided by the EPA, 2019.

Recruitment

Generally, the EPA does not feel it has difficulty attracting talent. Twenty-seven recruitment campaigns were run during 2018. Over 1 200 applications were received, of which 1 030 were from external applicants and the remainder from EPA internal staff.

The EPA has a policy of open recruitment and maintains itself as an equal opportunities employer. The EPA considers its workforce to be diverse, and the recruitment model allows for access for disabled groups. The EPA does not have a specific gender policy for recruitment, but rather relies on its equal opportunities policy to promote opportunities for women.

All roles are publicly advertised but the final decision on staff appointments following the competition are not published outside of the EPA. Job descriptions are created based on a skills model and competency framework, which identifies four core competencies that prospective employees must demonstrate experience and achievements: team player, communication, customer/stakeholder focus, and concern for quality and clarity of work. A review of competencies was completed in 2019 and will be implemented in 2020. The revised competencies are:

- Customer/Stakeholder Focus
- Interpersonal & Communication Skills
- Delivery of Results
- Analysis & Decision Making
- Team working/Leadership
- Specialist Knowledge & Expertise/Self Development.

There are no post-employment restrictions in place, nor is there any cooling off period following employment for any staff member. However, there is a cooling off period for directors who need to advise and seek approval of the Minister if there is a potential conflict. Any restrictions for staff are governed by the Staff Code of Conduct and, for directors, within their contracts. In addition, Ethics in Public Office declarations must be completed by all Staff.

Remuneration

The EPA follows central government salary rates, as sanctioned by DPER, and are at parity with similar positions in other government agencies. Staff are not eligible for additional benefits or exceptions for performance-based pay.

Talent retention and training

Turnover is very low at the EPA (see Table 2.6). However, staff retention rates of lower grades in Dublin is more difficult due to higher costs of living, longer commute times and opportunities for other employment due to higher economic activity. Of the 32 staff who resigned from the EPA between 2016 and May 2019, 22 moved to government or state agencies, six took up employment in the private sector and the remainder resigned while on career break.

The annual Performance Management and Development System (PMDS) is used to identify development opportunities with the staff member. The objectives of the PMDS are to ensure alignment of the performance and development of all staff with organisational goals and strategies. Staff and their line managers identify the learning and development actions required for effective performance in their current roles and to enable future career development, such as essential management and leadership programmes. The formal recording of this information is done through the PMDS.

The EPA also has a lateral mobility policy, which is intended to develop skills and fill vacancies. In general, EPA staff are assigned upon appointment but may be transferred to new assignments for organisational and/or development purposes. Managers review individual work assignments periodically to determine if an individual should be transferred to a new work assignment. This is ideally accomplished through the PMDS system, except for when new work arises between PMDS evaluations where the new work is incorporated into subsequent PMDS evaluations. The programme has two streams:

- Voluntary lateral moves: allows employees to apply for jobs via the PMDS system, which requires the manager to support the application and then the employee is placed into a central list for consideration.
- Management-initiated lateral moves: Are intended for employees who have been in posts for several years and are nominated for movement by HR, programme managers or directors to develop their skills in other areas. In some cases, this system has been criticised as being abrupt or lacking consent by the employee in question, as well as being poorly communicated. However, these moves have also been seen to be of benefit to staff development and to the organisation. Management initiated lateral moves cannot be between regions or grade levels.

While core competencies are identified for recruitment purposes, these competencies are not used for the life cycle of the employee in regards to determining probation, learning and development, promotion, development or training.

An annual Learning and Development Plan is approved by the Board. Its purpose is to link learning and development activities systematically with business needs and to establish priorities and plans for activities and resources. The development needs of the employee are also taken in to account, identified through the PMDS. The EPA allocates around EUR 600 000 per year to learning and development.

EPA is pursuing a "Keep Well" workplace wellbeing accreditation as a process to help the EPA achieve and sustain standards in workplace health, safety and wellbeing.

Performance assessment

The EPA has an online PMDS in place. The manager and employee set personal objectives at the beginning of the year, as well as when people change roles. These can sometimes happen months after the fact. A mid-year review is conducted, followed by an annual review at the end of the year. Objectives are supposed to be linked to strategic actions, as indicated in the strategic action plan. This link is stronger at the senior level than it is at the more technical level.

The PMDS does not rank the staff member, nor does it provide a bonus for staff members as the EPA is not permitted to provide bonuses. Salary step increases are also not linked to the PMDS system.

PMDS is intended to provide an opportunity for upward feedback as part of the interim and year-end review and constructive discussion is encouraged. However, some questions have been raised about the consistency in which upward feedback is utilised. A new staff engagement survey was conducted in 2018 as part of the HRD strategy.

Process

Decision making and governance structure

The EPA is managed by a full-time Executive Board that fulfils both management and strategic duties. The Executive Board comprises the Director General and five directors. The Director General serves as Chair of the Board and operational chief executive of the EPA. Each director also leads an office and provides day-to-day oversight of the EPA. Legislatively, the Board has responsibility for the management of the EPA but it is empowered to delegate responsibility to other staff for operational purposes. Currently twelve programme managers are delegated operational responsibility for carrying out the work of the EPA. The Corporate Governance manual gives guidance on this issue stating: "It is essential that there is a clear understanding of the role of the senior managers and the role of Board members – there is often a fine line between the two. The role of the Board is to approve strategies, policies and plans for the organisation and to monitor and review performance. It is the role of the senior managers to implement those strategies, policies and plans." The Board is appointed by the Government and accountable to the Oireachtas for the implementation of policy.

As defined in the Corporate Governance manual, the role of the Board is to provide strategic leadership, direction, support and guidance and promote commitment to EPA core values, policies and objectives. In addition to the special Board responsibilities set out in the *Code of Practice for the Governance of State Bodies* (DPER, 2016[4]) and in the EPA Act, the Board holds specific governance and management responsibilities as the Board of a State body which include:

- to ensure that the body carries out its responsibilities as set out by statute or by ministerial order;
- to define the mission of the body, decide its strategic goals and develop the policies required to achieve those goals;
- to ensure good management, to monitor the achievements of management and to ensure that a proper balance is achieved between the respective roles of board and management;
- to set performance targets, including key financial targets and, in particular, to agree and closely monitor the budget;
- to ensure that the body behaves ethically and in a manner that accords with the core values of the body; and
- to define and promote the body's role in the community by developing mechanisms for gathering the views of customers and stakeholders and by keeping people informed in an open, accountable and responsible way.

The Board has a set of decisions reserved to it to meet with its governance responsibility for the direction and control of the EPA, in compliance with Section 1.7 of the *Code of Practice for the Governance of State Bodies* (DPER, 2016[4]) which states that such functions should include the following:

- significant acquisitions, disposals and retirement of assets of the State body or its subsidiaries; the schedule should specify clear quantitative thresholds for contracts above which Board approval is required;
- major investments and capital projects;
- delegated authority levels, treasury policy and risk management policies;
- approval of terms of major contracts;
- assurances of compliance with statutory and administrative requirements in relation to the approval of the number, grading, and conditions of appointment of all staff;
- approval of annual budgets and corporate plans; and
- approval of annual reports and financial statements.

The Code stipulates the requirement for a Board Secretary. The Code's 2016 update expands the functions of the Secretary beyond secretariat duties to include reporting to the Director General on all governance matters and assisting in ensuring relevant information is made available to the Board and its committees. The EPA has assigned the roles and responsibilities of a Board Secretary to two individuals, as follows:

- Statutory duties, duty to exercise due care, skill and diligence, and duty of consulting the Executive Board through the Director General in all matters of governance: Programme Manager of the Corporate Governance Unit
- Duty of disclosure and administrative duties: Board Secretary

Legislatively, the Board has responsibility for the management of the EPA, but for practical purposes it is empowered to delegate responsibility to other staff for operational purposes. Section 25(6) of the EPA Act provides that the EPA may perform or exercise any of its functions through or by any director or other person or body who has been duly authorised by the EPA in that behalf. The twelve Programme Managers in charge of various functions are delegated operational responsibility for carrying out the work of the EPA. The Board delegates discretionary powers to various levels in the EPA. The delegation of powers has continued to grow as the EPA has acquired legislation over time. These powers have now been consolidated, are reviewed on an annual basis and are available on the EPA's intranet page.

The Board meets on a weekly basis. Meetings are classed either as General Board Meetings to consider organisational matters (once per month) or Technical Board Meetings (three times per month). General Board meetings cover items that are non-technical in nature such as the work programme; office operational reports; corporate governance items (e.g. internal audits; finance; human resources; communications; IT; organisational services); and strategic (e.g. organisational strategy; communications strategy). Technical Board meetings consider all items of a technical nature including all significant licensing and enforcement matters (e.g. approval of licences and legal Actions); other technical/scientific matters such as climate change and emission trading; environmental research; national waste prevention programme; legal actions; strategic environmental assessment update; EPA reports; etc. For both technical and general decisions, the relevant professional staff may attend meetings in order to answer queries and questions to inform the board in its decision making. On occasion the Board will discuss items amongst members only, depending on the nature of the topic.

Some licensing and enforcement decisions are delegated to director-level. Directors take the decision based on the submission made by the inspector or programme manager and records of these decisions are submitted for noting to a General Board Meeting. Legislation prohibits licence decisions below director-level. More complex licensing/enforcement decisions in terms of scale or technical complexity are dealt with by the Board during Technical Board meetings.

Guidelines for Board procedures are set out in the corporate governance manual, board meeting guidelines, procedures and timelines for Board papers, and Board meeting standing orders. The standing orders (guided by Sections 25 and 26 of the EPA Act) provide the set of formalised rules for the conduct of Board meetings, covering: organisation of Board meetings (location, timing, notice, Secretary responsibilities, validity); the quorum, set at three (although directors may decide to suspend Standing Order No. 8 to allow for the quorum to be set at not less than two, in accordance with Section 26(1) of the EPA Act 1992); lists decisions requiring a resolution of the EPA and EPA seal; minutes of meetings; and procedural decisions. Any of the standing orders may be suspended, amended or added to at any meeting provided that a majority of directors vote in favour. Papers are required in advance for all Board items for decision or for noting and must be submitted in a particular format and within pre-defined timelines.

The EPA Act 1992 states that every question at a meeting of the EPA Board shall be determined by a majority of votes of the directors present and, in the event that voting is equally divided and there are more than two directors present, the person chairing the meeting (usually the Director General) shall have a casting vote. In practice, decision making is by consensus rather than by vote and in the last fourteen years no vote has been taken by the Board.

Minutes are produced for all Board meetings. Extracts relating to licensing or enforcement issues are circulated to licensing and enforcement staff to be placed on the public file following approval by the directors. Extracts relating to all other issues are circulated to the author(s) of the Board paper following approval by the directors. The Director of OCCS circulates a 'highlights' report of Board meetings to all staff every two months in the internal staff newsletter. According to the Standing Order, the EPA may decide that any matter at a meeting will be confidential and not for public comment, unless specifically approved by the Chairperson. In the case of licensing issues, the Board Secretary may consult with a director before issuing extracts of minutes.

On occasion, the Board may discuss topics in an open way outside of official Board meetings, for example, discussing strategic challenges within a particular office or approaches to ICT management, as an opportunity for discussion prior to or separate from formal decision by the Board.

The Board has established the following committees to deal with specific issues:

- Audit & Risk Committee;
- Executive Risk Committee;
- Safety, Health & Welfare Board Sub-Committee, set up in December 2012 to provide leadership and visibility in relation to safety, health and welfare at work;
- ICT Board Sub-Committee, established in August 2015 to oversee the governance of all information management and technology related work and also to oversee the deployment of all staff and contractors working in this area;
- Ad hoc committees, as required, that are project work groups established by the Board for specific purposes. Normally, these groups report to a director in his or her operational role.

The Audit and Risk Committee (ARC) consists primarily of external members and is externally chaired. The role of the ARC is to provide independent assurance to the Board on the effectiveness of the control environment, risk management and the internal audit function. The Chair of the ARC attends the Board of the EPA at least once per year and prepares an annual independent report which is presented to the Board.

The EPA is advised by a number of other external committees, the primary such body being the Advisory Committee. The EPA is also advised by the following external committees: GMO Advisory Committee, National Waste Prevention Committee, Radiological Protection Advisory Committee, Dumping at Sea Advisory Committee and the Health Advisory Committee.

Internal organisation and management

The work of the EPA is divided into five offices, each reporting to a director:

- Office of Environmental Sustainability (OES)
- Office of Environmental Enforcement (OEE)
- Office of Evidence and Assessment (OEA)
- Office of Radiation Protection and Environmental Monitoring (ORM)
- Office of Communications and Corporate Services (OCCS)

In addition to the director, each office has two or three programme managers responsible for the implementation of the work programme of that office (Figure 2.4).

Figure 2.4. EPA organisational structure

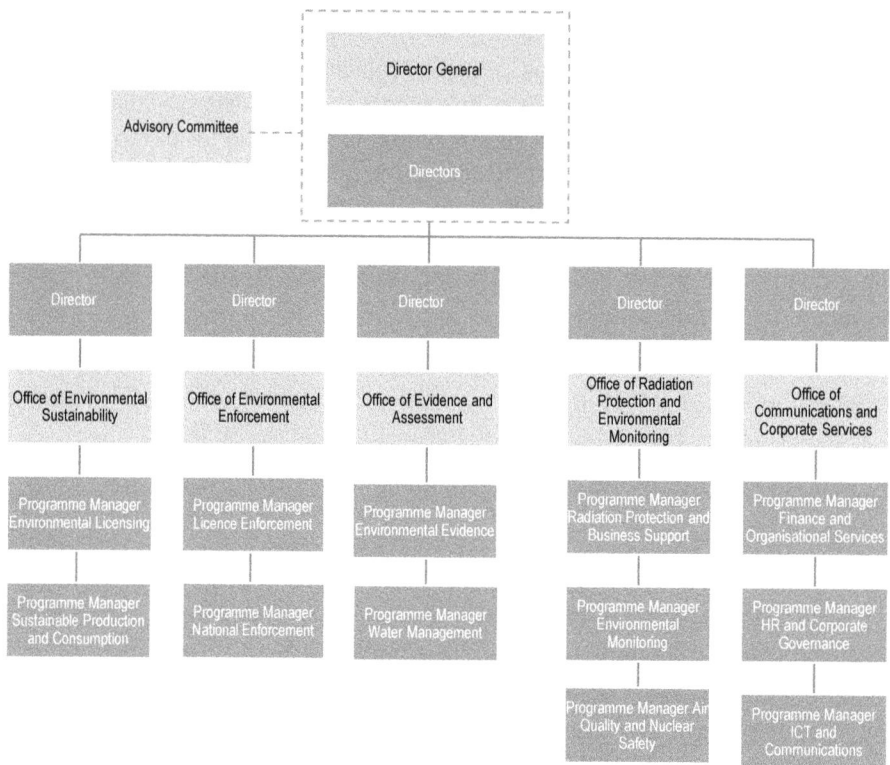

Source: EPA website, https://www.epa.ie/about/org/.

Functions are split across offices, for example: licensing activities are carried out by ORM and OES; enforcement activities by ORM, OEE and OES; data compilation and analysis by OEA, OEE, ORM and OES; advocacy activities are undertaken by all offices.

The Office of Environmental Sustainability (OES) has 81 staff and is divided into two programmes on environmental licensing and sustainable production and consumption that are responsible for licensing and some enforcement, compiling and reporting of national statistics, and advocacy functions.

- The environmental licensing programme is divided into six areas, covering: i) EU emissions trading; ii) Waste water/IE/IPC/Waste/Dumping at Sea iii) IE/IPC/Waste iv) Circular Economy & GMOs; v) Future legislation and EIA; vi) Administrative, support and improvement projects. In addition to these licensing functions, the EU emissions trading and GMO teams also carry out enforcement activities.

- The Sustainable Production and Consumption Programme carries out some activities in the area of climate change (compiling national data on greenhouse gas emissions and projections) and a broad range of activities in the areas of resource efficiency including: compiling national data for waste; leading the National Waste Prevention Programme; enforcing producer responsibility schemes for tyres, batteries and electronic equipment; and applying behavioural interventions, for example for food waste prevention. In addition, a small team within the programme provide a chemical regulatory service that includes enforcement functions.

The Office of Environmental Enforcement (OEE) counts 98 staff divided between five locations. OEE's work is organised into a national enforcement programme and a licence enforcement programme. Across both programmes, the OEE investigates failures to meet quality standards, prosecutes for significant failures, and produces guidance on best practice.

- The licence enforcement programme regulates large industrial and waste sites, dumping at sea and VOC permits through compliance promotion, inspection, monitoring and enforcement activities. Licence enforcement is organised into five regional enforcement teams, a financial provision and waste team, and an air team.
- The national enforcement programme regulates quality for drinking water and wastewater and produces the EPA annual reports on drinking water and wastewater quality. This programme also carries out the dual role of assisting and supervising local authorities' environmental performance. It advises and assists local authorities through its Network for Ireland's Environmental Compliance and Enforcement (NIECE). It also investigates complaints from the public about local authorities. The EPA's legal team is located in the national enforcement programme.

OEE organises conferences on topics under its remit. For example, the Water, Waste and Air conferences in 2018 were organised by the OEE with significant input and participation from OEA and ORM.

The Office of Evidence and Assessment (OEA) has 75 staff across seven locations divided between two programmes: evidence and assessment and water management. OEA provides the core of the EPA's knowledge functions and is also active in advocacy and public information.

- Evidence and assessment programme: Many of this programme's functions are derived from the EPA Act 1992, including the state of the environment reporting, co-ordinating Ireland's environmental research and liaising with the European Environment Agency. Other functions, such as the EPA role in strategic environmental assessment, were introduced later in line with new EU directives. Climate services is one of the newest functions housed in OEA. The office provides the secretariat for the National Climate Change Advisory Council and the National Dialogue on Climate Action as well as scientific advice to support Ireland's engagement in international climate negotiations. The analytics team within the evidence and assessment programme provides services to the EPA as a whole, using data science and visualisation to facilitate the work of different teams across the organisation.
- Water management programme: This programme provides national co-ordination and technical oversight for the Water Framework Directive (WFD) that includes: co-ordinating and engaging with around 50 organisations including state bodies and local authorities; managing the national WFD monitoring and reporting programme; and monitoring and assessing water bodies at over 4 000 sites (rivers, lakes, estuaries, coastal waters…) for WFD and national purposes. The water management programme also carries out hydrometrics monitoring and manages the National Hydrometrics Programme. In 2014, the catchment science and management team was added to the programme to develop and manage the scientific evidence base and integrated assessment tools used for river basin management planning and WFD reporting. Finally, the programme is

- responsible for reporting and communicating to the public on the quality of bathing waters and overall water quality in Ireland.
- As part of its advocacy functions, the OEA organises conferences, public lectures and workshops, sometimes in conjunction with other offices(e.g. conferences on climate change, health and environment, and water), and manages a number of dedicated websites – such as https://www.beaches.ie/ for bathing water quality information and catchments.ie which provides information and data on Ireland's 46 water catchments and Ireland's Environment, the web portal for information and indicators on the state of Ireland's environment.

The Office of Radiation Protection and Environmental Monitoring (ORM) was created in 2014 upon the merger of the Radiological Protection Institute of Ireland (RPII) and the EPA. ORM is divided into three programmes – air quality and emergency preparedness, radiation protection and environmental monitoring – which between them span all three core EPA functions of regulation (licensing and enforcement), knowledge and advocacy. Its 82 staff are distributed across six locations, including four water laboratories in Kilkenny, Dublin, Monaghan and Castlebar.

- The radiation programme is responsible for regulating the use of ionising radiation in industry and medicine (licensing and enforcement), for regulating occupational exposure to natural radioactivity and for providing a number of radiation protection support functions, such as a national dose register and approval or radon services and product certification. The programme also advises the government and informs the general public on non-ionising radiation.
- The air quality and emergency preparedness programme has five main areas of work. It is responsible for ambient air quality monitoring, modelling and forecasting and engages with the public around this theme by providing information and supporting citizen science initiatives. It leads the EPA's involvement as a key agency in the National Radon Control Strategy, provides advice and support to the government and local authorities and runs public information campaigns on radon. It leads the EPA's involvement as a key agency in the National Emergency Plan for Nuclear Accidents (acting as the national competent authority and providing technical support) and has more recently taken on a national support role for environmental emergencies. It advises the government on nuclear safety and supports the government in complying with international obligations. This programme also oversees the EPA's "citizen science" activities covering the themes of clean air, clean water and sustainability.
- The environmental monitoring programme oversees radiation monitoring and research carried out through the EPA funded research programme and water monitoring. The four water laboratories provide the physico-chemical monitoring required by the WFD and national programmes (the reporting of which is shared with OEA) and support the OEE by monitoring water at EPA licensed facilities, auditing waste water treatment plants, and supporting OEE investigations.

The Office of Communications and Corporate Services (OCCS) is organised into three programmes covering human resources and corporate governance (12 staff), ICT and communications (21 staff), and finance and organisational services (21 staff).

The EPA's headquarters is located in Wexford (150 staff) and it operates five regional inspectorates located in Castlebar (30), Cork (50), Dublin (120), Kilkenny (20) and Monaghan (14) and two smaller offices located in Athlone (2) and Limerick (2). Some national programmes are led from regional offices, for example, strategic environmental assessments are led out of Cork.

Structures have been put in place to improve co-ordination and communication between offices and regions. Given the decentralised nature of the EPA, a perennial challenge to meetings is the travel time and distances between sites.

- A senior management network (SMN) comprising directors and programme managers meet every two months. A key function of the SMN is the development and implementation of the EPA's strategy. The SMN has been running for three years examined in (McDonagh, Burke and O'Leary, 2018[7]). On occasion regional managers also attend.
- A leadership network was established in 2019 for middle-level managers (Levels 2 and 3).
- Meetings of technical functions take place three to four times per year and convene staff from regional offices.
- Cross-office groups convene staff to discuss particular topics, for example climate, waste, air quality working groups.

There have been a number of reviews of the EPA carried out in the last ten years:

- 2011 Review of the Environmental Protection Agency (EPA Review Group, 2011[8]) by the Environmental Protection Agency Review Group, an independent expert group appointed by the Minister for the Environment
- 2015 strategic review of EPA's regional presence
- 2015-16 External communications audit
- 2015 IAEA Integrated Regulatory Review Service Mission to Ireland
- 2011-2018 several value for money reviews on EPA programmes:
    - Environmental technologies and cleaner production research programme (2011)
    - The enforcement of the European Communities (Drinking Water) (No. 2) Regulations, 2007 (2011)
    - Water Framework Directive (2014)
    - National Waste Prevention Programme (2014)
    - Air enforcement activities (2015)
    - Learning & development programme (external review, 2016)
    - IMT support and maintenance arrangements (external review, 2017)
- 2018 mid-term review of Strategic Plan 2016-2020
- 2018 independent review of the National Waste Prevention Programme
- Annual reviews of Executive Board effectiveness
- Reviews or audits carried out as part of the Audit Plan, in additional to standard annual auditing programme:
    - Internal Audit of IM&T Security (2013)
    - Internal Audit of Licensing Activities (2014)
    - Review of Kerdiffstown Remediation Project (2014)
    - Review Management of Programmes funded by the DOECLG (2015)
    - External Review of Compliance with the 2016 *Code of Practice for the Governance of State Bodies* (2017)
    - Strategic Review of Emergency Response/Emergency Arrangements in the EPA (2017)
    - Internal Audit of Fixed Assets (2018)
    - Review of Procurement in ICT (2018)
    - Review of Payroll and Pensions in the EPA (2018)
    - Review of EU-ETS Processes & Procedures (2019, in progress)
- 2018 Independent review of the merger of the EPA – RPII by the Institute of Public Administration (IPA, 2018[9])

- 2019 OECD Performance Assessment Framework for Economic Regulators

In accordance with Section 7.3 of the *Code of Practice for the Governance of State Bodies* (DPER, 2016[4]), the EPA Board is responsible for ensuring that effective systems of internal control are instituted and implemented. This includes requirements for systems including financial, operational and compliance and risk management. The EPA Corporate Governance manual identifies several systems and procedures for internal control. Updates to the *Code* in 2016 placed a heightened focus on risk management and requires all state bodies to have an Audit and Risk Committee (the Audit Committee under the previous Code). Audits are conducted four to five times per year, which generally find that the level of compliance and control in the EPA is regarded as very high. The Executive Risk Committee (ERC) is responsible for further internal control through a Corporate Risk Register which is currently being updated to take into account the probability and impact of risks. Various other bodies further contribute to internal controls, including compliance with corporate legislation through a Compliance Officer's Report, ICT Compliance Report and Health and Safety Compliance report, and financial management and assurances.

Regulatory activities

The core regulatory function performed by the EPA is in regards to licensing/permitting and enforcement in various aspects of the environment sector. Originally, licensing and enforcement were together at the EPA but were separated in 2003 with the creation of the Office of Environmental Enforcement. The intention was to ring-fence resources allocated to enforcement and prevent what was seen as a conflict between competing priorities.

Currently, three of five internal divisions are involved in various aspects of licensing and enforcement: OEE, OES and ORM (see Table 2.7). For some sectors, OES carries out both licensing and enforcement functions. OEE only carries out enforcement in the areas of its competency. When the EPA was merged with the Radiological Protection Institute of Ireland (RPII) in 2014, licensing and enforcement for the radiological sector was maintained with the newly-created ORM.

Table 2.7. Responsibilities for licensing/permitting and enforcement at the EPA

Category	Licensing/permitting responsibility	Enforcement responsibility
Waste facilities	OES	OEE
Large-scale industrial activities	OES	OEE
CO_2 emissions trading	OES	OES
Intensive agriculture	OES	OEE
Genetically Modified Organisms (GMOs)	OES	OES
Drinking water by public water suppliers	N/A	OEE
Waste water discharges	OES	OEE
Dumping at sea	OES	OEE
Sources of ionising radiation	ORM	ORM
Large petrol storage facilities	OES	OEE
Local authorities	N/A	OEE
Waste Electrical and Electronic Equipment (WEEE)	OES	OES
Chemicals	OES	OES
Volatile Organic Compounds (VOC) Permits	OES	OEE
Air quality	Registration with EPA	Local authorities

Licensing and permitting

Licences and permits are granted by the EPA across a range of sectors (see section on Roles and objectives). On average, the process to issue a licence takes 1.5 years. Every licence is unique and requires detailed analysis by inspectors and a large breadth of information to make adequate decisions. The EPA aims to reduce this down to nine months.

The EPA runs a transparent process whereby all application documents, submissions and related information are available to the public on the EPA website. When gaps are noted in the application, requests for further information is made to the applicant and a notice is posted on the website. EPA inspectors are responsible for conducting a scientific assessment of applications and any submissions. This can include site visits to collect data. The licence proposal includes a detailed inspectors report and a comments matrix received with responses from the EPA.

All licenses go to the Board or are delegated by the Board to a Director for decision. The proposed licence or permit is followed by a statutory objection period, whereby the licensee can object to the proposed conditions. Third parties can also object to the proposed conditions. There is a fee at this point of EUR 250 to cover administrative costs. The technical committee reviews the objections and an oral hearing can also be requested by third parties. The licence can then be appealed by judicial review. However, this seems to be rarely used: of the 532 licence decisions taken from 2016-18, only eight have been appealed to Judicial Review. One of these has been ruled in favour of the applicant, five are still ongoing and the other two were either upheld or withdrawn.

According to Section 5 of the EPA Act, the EPA can issue a Best Available Technique (BAT) for licensees to follow, which seeks to use the most effective and advanced activity and method of operation to achieve a high general level of protection for the environment. BATs then become the basis for licence or permit approval. According to the EPA website (http://www.epa.ie/pubs/advice/bat/), there are currently 50 BATs in effect. Both the Waste Directive and Industrial Emissions Directive (IED) can impose BATs, requiring the EPA to update licences. When a BAT comes into effect, the EPA posts these to its website and must ensure all permit and licence conditions are updated, where necessary, within four years to be in alignment with the new provisions. These are generally accomplished through full reviews.

Revisions to licences can be requested when the regulated entity makes significant changes to their operation or new directives are imposed that change the licence requirements. According to the EPA Guidance for Licensees on Requests for Alterations to the Installation/Facility (EPA, 2019[10]), there are three pathways under which the EPA can decide to revise a licence: 1) where a change requires approval, but does not require a change in any conditions; 2) where the screening process indicates that the alteration is likely to require a technical or clerical amendment; and 3) where the screening process indicates that the alteration is likely to require a licence review or new application. The screening process is elaborated in the Guidance, which is designed to assist licensees in understanding and selecting the most appropriate mechanisms to their online request. However, it does not cover technical or clerical amendments or reviews initiated by the EPA. The guidance first poses 16 questions; a positive answer to any indicates that the change is significant and that the licensee should pursue a licence review.

New responsibilities to license have placed stress on the EPA to deliver licences efficiently. The Industrial Emissions Directive (2010/75/EU), which came into force in Ireland in April 2013, added the requirement to license large combustion plants. This resulted in the requirement for the EPA to license 50 combustion sites, causing a backlog. New requirements under the WFD will require the EPA to license larger water abstractions. Estimates forecast that this will require 800 new licences.

Enforcement

The OEE is responsible for most enforcement activities at the EPA for licences granted by the OES (see Table 2.8). Other areas where the EPA carries out enforcement activities (such as drinking water or producer responsibility schemes) are not required to hold licences. The OEE also supports the ORM in inspecting and enforcing EPA licensed facilities in the radiological sector. A number of indicators are tracked and published in the *Industrial and Waste Licence Enforcement Report* (EPA, 2017[11]) (see Table 2.9). Drinking water, wastewater, domestic wastewater treatment and local authority performance are all reported on annually, as is ionising radiation protection. According to the 2017 *Industrial and Waste Licence Enforcement Report*, the EPA's overall enforcement strategy is compliance-focused (Figure 2.5) and underpinned by the principles of:

- Proportionality in the application of environmental law and in securing compliance;
- Consistency of approach;
- Transparency about how the EPA operates;
- Targeting of enforcement actions where needed; and
- Implementation of the polluter pays principle.

Table 2.8. Types and number of licenses issued by the EPA

Area	Number of licenses
Industrial and waste	806
Wastewater	1 072
Dumping at Sea	14
VOCs	14
Radiological	1 749
GMOs	65
Emissions trading – Stationary installations	101
Emissions trading – Aviation operators	14
ODS End Users	35
Gas serving engineers/gas distributors	550
TOTAL	4 420

Source: Information provided by the EPA, 2019.

Table 2.9. Overview of indicators for enforcement activities in industrial and waste sectors, 2015-17

Category	2015	2016	2017
Number of non-compliances	1 612	1 546	1 619
Operational sites with no non-compliances recorded	58%	45%	54%
New compliance investigations opened	184	124	93
Number of site visits conducted	1 306	1 552	1 522
Percent of licensed sites visited	70%	72%	75%
Percent of licensed sites visited 2x	37%	42%	44%
Percent of licensed sites visited 3x	22%	22%	23%
Prosecutions	13	17	22
Total fines from prosecutions (EUR)	20 151 000*	179 000	375 000
Complaints received	1 031	1 101	1 030
Percent of complaints for odour	71%	66%	40%
Percent of complaints for noise	19%	18%	33%
Percent of complaints for air quality	4%	4%	20%

* In 2015, a landfill in Kerdiffstown was fined EUR 20 million by the Dublin Circuit Court.
Source: (EPA, 2017[11]), 2017 EPA Industrial and Waste Licensing Enforcement, https://www.epa.ie/pubs/reports/enforcement/EPA_Industrial_Waste_LE_Report2017.pdf; (EPA, 2016[12]), 2016 EPA Industrial and Waste Licensing Enforcement, https://www.epa.ie/pubs/reports/enforcement/EPA_industrial_waste_licence_enforcementReport2016.pdf; (EPA, 2015[13]), 2015 EPA Industrial and Waste Licensing Enforcement, http://www.epa.ie/pubs/reports/enforcement/EPAIndustrialandWasteLicenceEnforcement2015.pdf.

Figure 2.5. Types of EPA enforcement actions

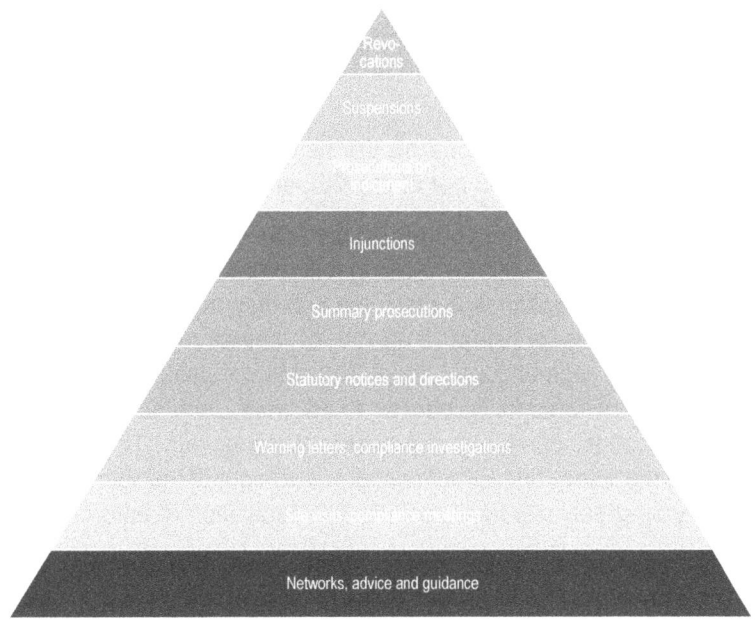

Source: (EPA, 2017[11]), 2017 EPA Industrial and Waste Licensing Enforcement, https://www.epa.ie/pubs/reports/enforcement/EPA_Industrial_Waste_LE_Report2017.pdf.

In September 2019, the Board approved the new Compliance and Enforcement Policy. The policy sets outs the high-level goals for compliance and enforcement with regards to enforcement principles, compliance actions, enforcement powers, criteria to consider when determining an enforcement action, and the communication of compliance and enforcement activities. The key principles underlying enforcement are stated as: risk-based, proportional, consistent, transparent and "polluter pays". The policy was developed in 2018, including a workshop in December 2018 to allow for cross-office input, and consulted on in April-May 2019 with stakeholders.

The EPA's risk-based approach categorises each site by the complexity of activity, type of activity, nature and quantity of emissions, sensitivity of receiving environment, and the location of the facility. The highest ranked facilities will receive the most planned visits each year, as well as visits in response to complaints from the public or environmental incidents at the facility. Each major sector (industrial and waste licences, drinking water and wastewater facilities, and radiation protection) has some individual variation in its risk-based approach.

An Annual Inspection Plan is generated through the Licensing, Enforcement, Monitoring and Assessment (LEMA) system, which uses a risk-based approach. This is integrated with the Environmental Data Exchange Network (EDEN), which provides an online portal for local authorities, Integrated Pollution Control (IPC), IED and Waste licensees to communicate with the EPA. The LEMA system was created in 2012 in an attempt to automate licensing and enforcement processes. The EPA notes that this system has allowed them to conduct better and faster reporting, reduce administrative burdens and led to more data-driven enforcement. LEMA allows the EPA to accept applications for industrial and waste licences electronically and engage with licensees during the subsequent enforcement stage online. Some 8 000 individual pieces of communication with licensees at the enforcement stage are now being done electronically and enforcement information can be done online for the benefit of the general public. Site visit plans by sector are as follows:

- For the industrial and waste sector, visits are based on the minimum requirements under the Industrial Emissions Directive (IED). The IED requires that the period between two site visits must be based on a systematic appraisal of environmental risks and must not exceed one year for high risk sites and three years for low risk sites.
- Drinking water and wastewater facilities are prioritised according to those facilities on the Remedial Action List and Priority Urban Area Action List. These lists identify higher risk facilities based on pre-defined EPA enforcement priorities.
- Radiation protection focuses inspections on the high risk practices, which is determined by the nature of the sector, practices carried out at the site, and past regulatory performance.

Inspections assess the licence holder's compliance with the licence conditions, which generally require the operator to conduct the licensed activity without causing risk or harm to human health or the environment. Detection and recording of a non-compliance is often the first step in escalation of enforcement action by the EPA. If the non-compliance is considered to be of environmental significance, a compliance investigation (CI) is opened. When opening a CI, the EPA inspector assigns a risk rating (high, medium or low) and may contain a number of items for a licensee to address, such as further monitoring, opening an investigation, providing clarification, carry out improvement works, etc. The CI remains open until the licensee has fully addressed the issue. The EPA maintains oversight of the licensee's progress in resolving each CI, and unsatisfactory progress may itself result in the escalation of enforcement action by the EPA.

The EPA has a range of compliance and enforcement tools available, some of which are specified in legislation and some are administrative tools designed to achieve compliance (see Figure 2.5). These range from advice and guidance to suspension or revocations of licences. While CIs are first used to drive compliance, failure to correct actions from a CI may result in a prosecution, which serves as the EPA's primary sanctioning power. Prosecutions at the summary level are subject to a maximum of EUR 5 000 fine and/or 12 months imprisonment. Files referred to the Office of the Director of Public Prosecutions (DPP) are subject to fines of up to EUR 5 million and/or 10 years imprisonment.

In limited circumstances, the EPA can give "on the spot" fines for some infractions, but in general it does not have recourse to this enforcement tool in other spheres. In 2018, legislation introduced a Fixed Payment Notice (FPN) system as part of enforcement for the WEEE and Batteries Regulations, which allows for "on the spot" fines of between EUR 500 and EUR 2 000 without recourse to court action. The process involves detection of a relevant non-compliance, issuance of a notification of non-compliance and, if non-compliance continues, issuance of a FPN. The agreed procedure requires that EPA send a recommendation to issue a FPN to Programme Manager, with final approval by the Director of OES. The process became live in 2019, with 12 FPNs currently in process pending final decision by the Director of OEE as of May 2019.

To further promote compliance, the EPA initiated the National Priority Sites system in 2017. The system ranks over 900 industrial and waste licensed sites according to enforcement factors such as complaints, incidents, compliance investigations, and non-compliances with the licence over the previous six months. Each metric is assigned points, arriving at a total score. Points for compliance investigations and incidents are assigned via a risk (high, medium, low) or outcome (catastrophic to minor) based categorisation. Points for complaints and non-compliances are tallies based on number of each infraction noted, though complaints only count for those linked to a medium or high compliance investigation. The system intends to drive compliance through behavioural change by "naming and shaming" the poorest performing sites. The list is updated quarterly. According to the 2017 *Industrial and Waste License Enforcement Report* (EPA, 2017[11]), 20 sites were identified as being part of the list in 2017. Of these sites, 14 stayed on the list one out of four quarters and only one site was on the list for all four quarters. In September 2019, the Board approved changes to the NPS that *inter alia* amends the ranking formula to reduce the weight of complaints.

The EPA also engages in compliance-promotion activities, including producing guidance, engaging with sector and industry groups, and circular letters to industry identifying priorities for the year. The EPA also works to identify the key risks and priorities for particular sectors and seeks to address them through a combination of compliance and direct enforcement.

An innovative programme to drive compliance is the Smart Farming initiative developed through a strategic partnership between the EPA and Irish Farmers Association (IFA). This voluntary on-farm resource efficiency programme aims to drive behaviour change by allowing farmers to develop their own strategy for reducing greenhouse gas emissions by 5-7% and making cost savings of EUR 5 000. While the EPA maintains the right to sanction farmers for non-compliances, the programme intends to partner the EPA and IFA to jointly develop solutions to promoting economic and environmental outcomes.

In 2014, the EPA introduced a policy requiring certain regulated entities to set aside financial provisions as part of a licence to pay for future site remediation and any potential clean-up that may be required. The current policy identifies 150 sites requiring financial provisions, with a total of EUR 850 million set aside to meet the requirements. The original policy from 2006 was updated in 2015 and provides updated guidance on the standards necessary to comply with relevant EPA licence and permit conditions. While previously financial provisions were allowable only by cash, the financial provision now acceptable to the EPA include: secured fund, on-demand performance bond, parent company guarantee, charge on property and environmental impairment liability insurance.

Co-ordination

Internally, some co-ordination occurs due to one office having responsibility to undertake activities in the domain of another office, but these instances are fairly limited. Moreover, where required, information is shared between offices, including site-specific information for health and safety purposes.

Externally, the EPA works through networks aimed at sharing information about enforcement. First, an Environmental Enforcement Network was established in 2004 to improve the overall level and consistency of environmental enforcement in Ireland. This network has since evolved into NIECE – the Network for Ireland's Environmental Compliance and Enforcement. All state bodies with a role in environmental enforcement and compliance are members of this network. Second, a similar collaborative networking approach has been adopted for improving the overall implementation of the Water Framework Directive, which is led by the Water Programme in OEA.

EPA liaises with a number of state bodies who also have inspections responsibilities. On occasion, the EPA has engaged in co-ordinated site visits with other state bodies, such as An Garda Síochána, the HSE and through the Local Authority Waters Programme (LAWPRO). However, this is not systematically organised.

Local authorities function with a dual role in relation to the EPA, in that they receive advice from the EPA as well as are subject to enforcement in their areas of competence. While this dual role is regarded as fit for purpose by the EPA and local authorities, there does require efforts to clarify roles and promote collaboration. WERLA and LAWPRO shared services are an attempt at solving some of these local issues. These shared services are established through a process overseen by the Local Government Management Board.

Waste Enforcement Regional Lead Authorities (WERLAs): DCCAE nominated three local authorities as the lead in their respective regions (Cork County Council for the southern region, Dublin City Council for the eastern and midlands region and Leitrim & Donegal County Councils (combined) for Connacht/Ulster

Regions) for co-ordinating enforcement actions, setting priorities and common objectives, and ensuring consistent enforcement of waste legislations while leaving local authority personnel as first responders to specific breaches of waste legislation. WERLAs are overseen by a National Steering Committee, which includes representatives from a wide range of regulatory authorities, including the EPA. National waste enforcement priorities are set by the committee and aim to drive consistency at a central level. Other stakeholders in the waste sector have an opportunity to input into this enhanced waste enforcement structure through an Industry Contact Group.

LAWPRO was expanded from the previous Local Authority Water and Communities Office in 2018 as part of the current River Basin Management Plan (RBMP) as required under the Water Framework Directive. The LAWPRO brings together all 31 local authorities to achieve common water quality goals. The LAWPRO seeks to protect and improve water quality by 1) support and co-ordinate public bodies and other stakeholders to achieve objectives in the RBMP; 2) Activate local communities to engage with river catchment in line with the integrated catchment management approach; and 3) Build a foundation and momentum for long-term improvements and inform the development and implementation of the 3rd Cycle RBMP. Coordination with LAWPRO is led out by the Water Programme in OEA.

Regulatory policy tools

Ex ante assessment of impacts

The EPA is one of a number of competent authorities in Ireland for conducting Environmental Impact Assessments (EIAs), mainly relating to activities licensed by the EPA. EIAs are required for planning permission or other development projects in regards to Integrated Pollution Control (IPC), Industrial Emissions (IE) and Waste license applications. As well, the EPA is competent authority for certain Waste Water Discharge License (WWDL) applications.

As of May 2017, EIAs must meet the requirements set forth by EU Directive 2014/52/EU, which is implemented in Ireland by DHPLG Circular letter PL 1/2017. The 2014 Directive places responsibility on the developer to prepare an Environmental Impact Assessment Report (EIAR) in areas of their competency, and responsibility to the EPA to provide reasoned conclusions following the examination of the EIAR and other relevant information.

The 2014 Directive requires that the EIAR must identify, describe and assess appropriately the direct and indirect significant effectives of the project on the environment. EPA produced the draft guidelines on EIARs using a risk-based approach to determining the significance of effects, which plots impacts in accordance with the probability and consequence of the impact. The EPA is empowered to have access to sufficient expertise to examine the EIAR and may seek supplementary information to reach a reasoned conclusion.

The draft guidelines for EIARs identify thresholds set out in legislation for determining if an EIAR is required. The only types of projects to which thresholds do not apply are types that are considered to always be likely to have significant effects; a crude oil refinery for example.

Cost-benefit analysis and economic appraisals

In addition to EIARs, license applications, reviews and enforcement decisions must be supported by an assessment by the EPA inspector with appropriate recommendations, which the Board uses to take a final decision. These include a consideration of:

- *Costs:* as potential costs to the environment, particularly externalities. However, there is no attempt to put an economic cost or evaluate the economic impacts but does attempt to balance the protection of the environment against the need for infrastructure, economic and social progress and development. Some licensees are required by the EPA (through the licence) to estimate the financial provision required to deal with closure, aftercare and environmental damage costs and to agree these costs with the EPA.

- *Benefits:* as potential benefits to the environment through the prevention of environmental pollution. The EPA is not obliged to quantify this benefit in financial, economic or societal terms as the legal requirement on the EPA is that it cannot grant a licence unless it is satisfied that the licensed activity will, amongst other things, not cause environmental pollution. However, if a licensee is seeking a derogation from BAT under the Industrial Emissions Directive, then the application for a derogation must be supported by a cost benefit analysis, for which the EPA has issued guidance on this (EPA, 2016[14]).

The EPA also uses a number of risk assessment tools for unauthorised waste activities, septic tanks, abandoned mines and WFD characterisations. Spatial analysis and environmental assessment tools are also used to support assessment and decision making functions. Many of these tools are publicly available through the EPA's GeoPortal (http://gis.epa.ie/).

Stakeholder engagement

The EPA adheres to the Aarhus Convention and engages in stakeholder engagement in two areas: 1) Evaluating licence applications and 2) developing new guidelines and processes.

When an industrial, waste or wastewater licence application is received, the request goes immediately online via the EPA website and is open for public consultation. Throughout the application process, multiple opportunities arise for stakeholders to provide comments (more details presented below). All comments are posted online.

In addition, for licences requiring an EIAR, the EPA must consult with the prescribed bodies before giving an opinion, which include the Health Services Executive (HSE), the Health and Safety Authority (HSA), and local authorities. Conversely, the EPA is a statutory consultee to An Bord Pleanála (ABP) and must make observations and submissions in relation to EIARs for local authority developments and strategic infrastructure developments that comprise IPC, IE or waste licensable activities. The EPA can also give advice on other matters when requested by the planning authority or ABP.

The process for obtaining statutory responses begins with a notification from the Planning Authorities or Local Authorities to the EPA in relation to proposed developments associated with IPC, IE and waste licensable activities. The EPA may then make submissions/observations on these notices.

The EPA responds to notices from ABP regarding a planning application for development comprising of a waste licensable activity. A Planning Authority or ABP may request the EPA to make observations in relation to a proposed development that, in its assessment, is likely to have a significant impact on waste water discharges. The EPA will make observations on the assessment undertaken.

ABP may also consult with the EPA on Strategic Infrastructure Development projects, which are developments considered of strategic economic or social importance to the State or region covered under the Planning and Development (Strategic Infrastructure) Act 2006. The consultation focuses on the transboundary aspects, whereby the EPA is asked for its opinion of an EIAR.

EIAR information must be made available to the public electronically and by public notices. The EPA must inform the public and prescribed bodies of its decision, as well as make available information on the content of the decision and any conditions attached. This includes the main reasons and considerations for the decisions, including information about stakeholder engagement and a summary of the results of consultations and information gathered during the EIA process.

When developing new guidelines and processes, the extent of consultation depends on the nature of the guidance. The level of significance of a new guideline or process is often determined based on whether the guidance is statutory or not. Some are produced under parent legislation and have legal standing whereas others are best practice.

For more significant guidance, the EPA will consult with stakeholders prior to the preparation of the guidance and may form a steering committee to aid the preparation of the guidance. Draft documents open for public consultation are posted to the EPA website, which is normally conducted two to three times per year (see Table 2.10). There are no internal requirements on how consultations are to be conducted. This includes length of consultation periods. If consultations take place via a steering committee, there is no statutory period for comments and the discussion will continue until a satisfactory outcome is achieved. Where the document is issued for public consultation via the website, it can be for varying periods; no minimum periods are prescribed by law. When the guidance is complete it is then issued to key stakeholders and in some cases the general public for consultation. For example, the preparation of the Code of Practice for Domestic Wastewater Treatment Systems followed this process.

Table 2.10. Consultative documents, 2014-18

2014	2015	2016	2017	2018
National Waste Prevention Plan 2014-2020	Draft Better Regulation Policy		Draft EPA Guidance on requests for alterations to a Dumping at Sea Permit	Guidance on Soil and Stone By-products
Proposed Guidance on the Authorisation of Direct Discharges to Groundwater	Draft revised Guidelines on information to be contained in Environmental Impact Statements; and Advice Notes for preparing Environmental Impacts		National Inspection Plan 2018-2021: Domestic Waste Water Treatment Systems Draft for Consultation	
EPA Viewpoint on the use of European Waste Catalogue code 19 12 09.	Preparation of a Draft Language Scheme by the EPA.		Draft Guidance Note on Soil Recovery Waste Acceptance Criteria	
Environmental Regulation of Healthcare Risk Waste Storage & Treatment				
Preliminary Consultation on Revisions to the Environmental Impact Statements Guidelines				

Note: Following public consultation, it was decided to not further pursue the Better Regulation Policy.
Source: http://www.epa.ie/pubs/consultation/.

Where the guidance is less significant the EPA will target specific stakeholders for consultation e.g. advice notes in drinking water are issued to a limited range of stakeholders prior to release and not issued for general consultation.

The EPA can also create ad hoc steering committees to provide input on the guidance itself which then disband once the work is complete. These are separate to the standing committees noted in Table 2.11. There are no set rules for how frequently they meet as it is specific to the guidance. For example, some will meet just once whereas others can have several meetings depending on the complexity of the guidance.

Committees and networks

According to the Corporate Governance Manual 2018, the EPA Board is advised by several external committees and groups that also serve as bodies that the EPA engages with in regards to its programme of work (Table 2.11).

Table 2.11. EPA external committees

Committee	Legal standing	Composition	Functions
Advisory committee	Required under Section 27 of the EPA Act	12 members: Chair: Director General of the EPA Seven members nominated by organisations concerned with environmental, development or wider social, economic or general matters Four members appointed by the Minister of DCCAE Term limit: 3 years Meeting frequency: 4 times per year	Make recommendations to the EPA or to the Minister related to the functions of the EPA, as dictated under Section 28 of the EPA Act. The Advisory Committee is not entitled to receive specific information in relation to the application or review of licences or provide any recommendations regarding licensing.
GMO Advisory Committee	Advice given under the GMO Contained Use and Deliberate Release Regulations	14 members, with Chair from the EPA. Members are nominated by government and non-government organisations. Term limit: 3 years Meeting frequency: Once per year	The GMO committee is a consultative body that advises the EPA on relevant GMO issues for consideration by the Board of the EPA where relevant.
National Waste Prevention Committee (NWPC)	Convened by the Minister in 2004. Section 74(13) allows for establishment of a committee for monitoring performance	20 members, with Chair and Secretariat support by the EPA. Members are drawn from government, non-governmental, business and sectoral interest groups. Term limit: Membership is for a three-year period renewable by appointment of DCCAE following proposals from EPA. Meeting frequency: At least twice per year where a quorum of 50% plus one is required	Monitor the development of the National Waste Prevention Programme and provide strategic direction for the EPA in implementing it.
Radiological Protection Advisory Committee	Established in 2016 following the merger of RPII with the EPA in relation to Section 41 of the EPA Act	16 members nominated by organisations with expertise relevant to the radiological protection functions of the EPA. Term limit: 3 years Meeting frequency: Twice per year	To act as a high-level scientific advisory body to advise the EPA in the carrying out of its functions on matters concerning radiological protection, with particular emphasis on public health.
Dumping at Sea Advisory Committee	The Dumping at Sea Advisory Committee was established in February 2010, under Section 41 of the EPA Acts 1992 to 2007	Two meetings of the Dumping at Sea Advisory Committee were held in 2017 and one in 2018. Further interactions between the EPA and the committee were conducted over the period, via electronic communications. In 2018, the committee was consulted on all permit applications received and an agency initiated amendment. Term limit: 3 years	To advise the EPA with respect to the administrative and technical implementation of its functions under the Dumping at Sea (DAS) Acts 1996 to 2010.

Committee	Legal standing	Composition	Functions
Health Advisory Committee	Established May 2012 in accordance with Section 41 the EPA Act.	The committee comprises representatives from public bodies working on environment and health issues, including the Health Service Executive, Health and Safety Authority, Health Research Board, Department of Communications Climate Action & Environment, Department of Agriculture Food and the Marine, Department of Health, An Bord Pleanála, Food Safety Authority of Ireland, Health Products Regulatory Authority, Department of Housing Planning and Local Government and the County and City Managers Association. The committee met three times during 2018	To assist and advise the EPA in relation to the public health implications of matters pertaining to environmental protection.

Membership in the external committees is without remuneration, except travel expenses to attend meetings. It is possible for committee members to attend meetings remotely via video conference. Meetings for some committees, such as the Advisory Committee, are regularly held in different locations to accommodate the regional presence of the EPA staff.

The Advisory Committee meets most often of all the external committees. The Chair is the Director General of the EPA, who sets the draft agenda for each committee meeting following input from the members. Secretarial support also provided by the EPA. Meetings are usually half day long, with information to be discussed given to committee members prior to the event and members are asked to provide their opinions, including on the strategic objectives of the EPA. Key priorities for discussion in the Advisory Committee are set for the year by the Chair in consultation with the members.

The EPA Act allows the Advisory Committee to provide recommendations to the Minister of DCCAE under Section 28, which lists what subjects the Advisory Committee may make recommendations about but does not establish any formal mechanisms. High-level reports are published at the end of the three-year mandate for the Advisory Committee. These reports are sent to the Minister of DCCAE, although there is no formal requirement to do so, and the format, scope and level of detail of the reports varies between committees. Minutes from the meetings are posted to the EPA website, with short summaries of discussions and broad action points. Decisions are made by consensus.

The 2011 Review of the EPA recommended making the Chair an external member. In 2014, the Department of the Environment, Community and Local Government contacted the EPA seeking its views, and the views of the Advisory Committee, on this recommendation. After discussion with the Advisory Committee, it was concluded that, on balance, an external chair would not benefit the committee at that present point in time and that a better approach would be to revisit the role and purpose of the Advisory Committee as it had been in operation for 20 years. It was agreed that the committee would assess how it currently operates, including how to optimise the committee's work and the formal facilitation of advice to the EPA and the Minister. The 2011 Review further recommended an emphasis on the selection of members with particular knowledge and experience of the environment, as well as stronger representation of key stakeholders, including those in the public sector, to better integrate the public service with the environment sector. Legislation was amended to facilitate this.

The EPA also engages with a number of networks, which includes business and NGO actors. One of which is the Irish Environment Network (IEN), which the EPA meets with on a biannual basis. The IEN consists of non-governmental organisations (NGOs), and meetings include staff representatives from the EPA's different teams, depending on the agenda. However, these networks and committees are not used to conduct early-stage consultations on EPA activities.

Appeals and complaints

Ultimately the appeal mechanism available to challenge a licensing decision by the EPA is a challenge by judicial review. Several interviewees referred to the high cost of litigation in Ireland and the challenge this poses in a context where environmental legislation is becoming increasingly complex.

Table 2.12. Appeals

Year	Number of final decisions taken across all licensing streams	Number of decisions appealed	Status (decision upheld, rejected, on-going)
2018	94	5 judicial reviews on Art. 27s	5 conceded
2017	108	1 judicial review on WWDL	Court ruled in favour of the applicant
2016	165	1 judicial review on waste 1 judicial review on Art. 27	Decision upheld. Appeal withdrawn

Citizens can make complaints about the service provided by the EPA in accordance with its Quality Customer Service Charter – see (EPA, 2019[15]). Citizens can make environmental complaints via several mechanisms – see (EPA, 2019[16]) including a dedicated telephone line, an EPA app (See It, Say It), email address and online forms.

Environmental complaints are received predominantly by phone, followed by written complaint using email, the *See It, Say It* app and online forms. A monthly summary of the complaints received and how they are managed by the EPA is produced for internal circulation and management of work. Environmental complaints are also received by the EPA that are directed to other responsible bodies such as local authorities. These are predominantly managed through the National Environmental Complaints Line (157 complaints in August 2018) and the See It Say It app (298 complaints in August 2018).

Non-regulatory approaches

The EPA regulatory responsibilities are focused on the delivery of government policy, while regulatory design is the responsibility of DCCAE and DHPLG. The EPA does encourage non-regulatory approaches to environmental issues where appropriate. This includes:

- Advocacy in areas of strategic importance to facilitate and encourage changing behaviours and attitudes towards the environment. The EPA has developed a suite of educational material for use in schools and communities (EPA, 2019[17]).
- Use of 'soft regulation' technique by way of networking and engagement activities. For example, collaborative structures have been created (and underpinned by legislation) at national, regional and local level to facilitate collaboration between all of the state bodies with a role in water protection and management with clear assignment of roles and responsibilities. This approach has been embedded in the National River Basin Management Plan 2018-2021.
- The National Waste Prevention Programme (EPA, 2019[18]) provides support and guidance to businesses, households and the public sector. Amongst its activities, the Programme partners with other relevant national organisations to advocate for increased resource efficiency for waste, water and energy. Some examples of its initiatives are Stop Food Waste and Smart Farming in partnership with the Irish Farming Association.
- National Dialogue on Climate Action, which is a new national programme with the EPA assigned as the co-ordinator. This is an ambitious programme aimed at engaging the general public about the issue of climate change and what Ireland can do to deal with both mitigation and adaptation.

Transparency, integrity and accountability

The EPA adheres to the Aarhus Convention (European Commission, 2019[19]), which establishes a number of rights of the public (individuals and their associations) with regards to the environment. This includes the right to have access to environmental information. As developed above, the process for licensing and engagement uses online portals for interaction with stakeholders.

Transparency *vis-à-vis* stakeholders is delivered primarily through the EPA website. The main page is organised with many links and sections that are broadly organised around the EPA's three functions of "regulation", "knowledge" and "advocacy". However this structure is not explicitly on display, nor are these functions listed on the EPA main page. Links and documents related to EPA processes and outputs of interest for stakeholders are partly on display, while others are located in hard-to-find locations.

The EPA adopted a policy to use plain language in 2017. In 2018, a workshop on the use of plain language was given at the EPA. The first to use such an approach was in the Drinking Water Report (EPA, 2018[20]) where the EPA adopted the National Adult Literacy Agency approach.

Standards of behaviour at the EPA are governed in legislation by the EPA Act, the *Code of Practice for the Governance of State Bodies* (DPER, 2016[4]), and the *Ethics in Public Office Acts* 1995-2001. No cases of staff or senior management being in violation of these acts has been documented. More generally, the 1995 Nolan Committee (United Kingdom) recommended seven principles of public life (selflessness, integrity, objectivity, accountability, openness, honesty and leadership) and are adopted at the EPA as standards underpinning the legislative standards for all directors and senior managers.

An EPA Code of Conduct is required by the *Code of Practice for the Governance of State Bodies*, which must include policies addressing integrity, information, obligations, loyalty, fairness, work and external environment and responsibilities. The EPA implements this requirement with the *Code of Business Conduct for Directors and Staff* that establishes general principles and standards to govern the professional activities and conduct of Directors and Staff of the EPA, with the goal of maintaining a high level of public confidence in the organisation as a public body and as an employer.

Declaration of interests is governed by Section 37 of the EPA act, which requires relevant parties – including the Director General, directors or other employees or persons – who have interest in any land or activity under scrutiny by the EPA to complete a Declaration of Interest Form to the EPA.

Disclosure of interests is governed by Section 28 of the EPA Act. The provision applies to the Director General, directors, any employee, Advisory Committee member, EPA committee or consultative group, consultant, advisor, or any other person engaged with the EPA who has a financial or other beneficial interest in any matter considered by the EPA. The policy requires disclosure in advance of considering the matter, a declaration that the person will not influence or seek to influence the decision, take no part in considering the matter, and withdraw from meetings.

Confidentiality is governed by Section 39 of the EPA Act, which prohibits Board members and staff from making use of or disclosing confidential information gained as a result of employment with the EPA. This provision continues even following the member leaving the EPA for any reason, and results in disciplinary action for violations. Special provisions under Section 32 of the EPA Act allow for Freedom of Information (FOI) requests.

Whistleblowing is governed under the *Protected Disclosures Policy and Procedures* in accordance with the *Protected Disclosures Act 2014*, which protects the identity of employees who make disclosures. The EPA has appointed a Protected Disclosures Officer (PDO) to deal with all protected disclosures. The PDO resides within the Corporate Governance team. The protected disclosures policy complements the EPA's Anti-Fraud and Anti-Corruption Policy, which commits the organisation to maintaining a culture that opposes irregularity, fraud and corruption.

A Dignity at Work policy, as well as a Grievance Policy and Procedures to address grievances are in place. The Dignity at Work policy covers bullying and discrimination, amongst other topics. Complaints are first filed with the line manager unless it is the line manager against whom the complaint is being made in which case the complaint is escalated to the next line manager up the chain of command. If a resolution is not found, it is escalated to the programme manager and then HR. The goal is to seek a resolution to the issue. There is not an explicit avenue for receiving complaints from women, which would not fall under the Protected Disclosure Policy. In the past four years, there has only been two complaints made by staff members and these were handled locally through a mediation process.

Accountability at the EPA is maintained through annual reports, regular reviews of policies and practices, research and policies (see section on performance reporting below). The EPA is accountable to the Oireachtas through the Public Accounts Committee.

Output and outcome

Data collection

The EPA collects data from regulated entities in the framework of its enforcement activities. These include:

- Industrial, waste and wastewater licensees: monitoring data and an Annual Environmental Report (AER) which provides a summary of emissions and environmental performance information.
- Irish Water: water quality monitoring data annually.
- Local authorities: Annual inspection and enforcement plans and results of previous annual plans. Local authorities all have different IT systems for capturing data.
- Radiological licensees: data as per their licences.

The EPA has developed an automated online tool to tackle shortcomings in data submissions for water and wastewater data, representing approximately 200 000 test results annually. This system automatically checks for errors, improving the quality of data received and reviewed by the EPA.

Data is used to identify key compliance issues per sector, for water, wastewater and for local authorities. In the latter case, given the EPA's role on providing support and strengthening local authorities' capacities, this analysis has allowed for the EPA to better target its actions. For example, analysis of the *E. coli* in drinking water data indicated that most of the failures were of a short duration and caused by temporary failures in treatment. A concerted enforcement effort was put in place to install disinfection monitors and alarms in all public water supplies contributing to a 90% reduction in incidents.

In 2018, the EPA established a small data analytics team to pilot the use of data science, spatial analysis, earth observation and data visualisation techniques, working in close collaboration with EPA subject matter experts. For example, working with the Urban Wastewater Treatment data that Irish Water submits to the EPA via EDEN, the analytics team used statistical methods to group the monitoring results into "improving", "staying the same" or "getting worse" for each of the different parameters to produce an Urban Waste Water Scorecard. This allows inspectors to quickly focus on the specific plants and parameters that are a problem among the thousands of data points.

The EPA also collects large quantities of data to monitor and assess Ireland's environment, fulfilling several statutory reporting duties to the national government and the EU (e.g. water quality monitoring for the WFD). The EPA provides near real time data on air quality and hydrometrics through its monitoring networks. The EPA manages Ireland's Environmental Open Data Portal that is primarily intended as a resource for software developers rather than key stakeholder groups or the general public. It gives access to data collected for the WFD as well as data on bathing water and radiation.

Monitoring and reporting on performance

Regulated entities

The EPA reports transparently on the performance of regulated entities in all of its areas of work. These include an annual review of the performance of these facilities is carried out and published in the annual reports on drinking water (EPA, 2018[20]), wastewater (EPA, 2017[21]) and regular updates on industrial and waste activities regulated by the EPA. In addition, all AER reports are published online.

The EPA also implements a successful naming and shaming strategy via the publication of a ranking of licensees and sites in the different sectors, update on a quarterly or annual basis. This strategy is claimed to have achieved good results as licensees and sites do not wish to appear on these lists:

- Facilities prosecuted by the EPA
- Drinking water treatment plants that are failing or at risk of failing to meet the required standards (see Remedial Action List at http://www.epa.ie/water/dw/ral/)
- Waste water treatment plants failing to meet the required standards or posing a risk to the environment (see Priority Urban Areas at https://gis.epa.ie/EPAMaps/SewageTreatment)
- Industrial and waste sites prioritised for enforcement based on environmental performance (see National Priority List at http://www.epa.ie/enforcement/nationalprioritysites/ - d.en.62512)

There are no events around the performance of sectors that convene regulated entities to discuss performance and compliance. Licensees are however sometimes invited to conferences, for example the National Air conference.

Ireland's environment

The EPA monitors and reports on Ireland's environmental quality through a series of publications, including its flagship "state of the environment" report, which is published every four years. The EPA publishes more frequent reports on specific environmental outcomes, such as air quality, water quality, waste and climate change. The findings of reports are usually promoted through press releases at the moment of their launch. Data and indicators on environmental performance in various sectors can also be found on the EPA website under the different issue areas (e.g. waste, air…); data and statistics are not compiled in a single place on the website.

The EPA

The EPA sets out five-year strategic plans. The plan for the current period 2016-2020 sets out five goals, 14 objectives and 16 outcomes to achieve by 2020. The five goals cover all areas of activity of the EPA and mainly focus on the EPA itself (regulator, leader, partner, organisationally excellent etc.) with one goal on responding to key environmental challenges. The 2016-2020 Strategic Action Plan assigns activities to each of the objectives. These documents do not include any quantitative targets or metrics.

Internally, the EPA strategic plan is translated into a yearly work programme that lists activities for each office. It is updated and presented to the Board once a month, in a report structured into sections per office, compiled by the Corporate Governance Unit. This monitoring focuses on the implementation of activities and projects.

Externally, the relationship between the EPA and DCCAE and the DHPLG is governed by a three-year oversight agreement (current period 2019-2020). Annex 2 (Performance/Service Levels) of Appendix 1 (Performance Delivery Agreement) of the oversight agreement lists functions, outputs, timeframe and performance indicators for regulatory functions of the EPA within the remit of DCCAE and DHPLG respectively. The current agreement includes over 100 indicators in total. The focus of these indicators is on meeting legislative obligations, producing annual reports and holding annual events; they do not seek to measure the quality of processes or outputs and do not include any outcome or impact level indicators.

In practice, the EPA does not report on all of these indicators to the departments; instead, in the case of the DCCAE, 8 indicators out of the total have been selected for reporting on an annual basis. These are summarised below in Table 2.13. The focus is on metrics (outputs) rather than quality of processes, outcomes of activities, or overall sector performance (i.e. water or air quality, safety of industrial sites…). These appear to be the only quantified targets that the EPA reports on, internally or externally.

Table 2.13. Indicators reported by EPA to DCCAE

DCCAE - EPA REV 2018	Output outturns					Output targets	
	2013	2014	2015	2016	2017	2018	2019
Key high-level metrics							
Number of Environmental and Radiological Decisions (Note 1)	2 469	2 100	1 885	3 414	1 610	2 232	2 140
Number of Industrial/Waste Site Visits (Note 4)	1 370	1 357	1 310	1 558	1 529	1 320	1 320
Number of Urban Wastewater and Drinking Water Site Visits	299	399	338	340	377	370	370
Legislation							
n/a (Note 2)							
Publish							
Number of EPA Reports published	27	25	34	38	49	35	35
Number of Reports on Environmental Research Projects published	29	21	25	35	38	35	35
Number of Open Data datasets on the DPER Open Portal (Note 3)	50	50	98	140	228	250	260
Context and impact indicators							
Number of visits to EPA website	750 000	780 000	800 000	819 000	908 000	860 000	900 000
Number of environmental queries from the public answered	2 700	2 650	2 500	2 207	2 184	2 200	2 200

1. Includes licences, certificates of registration, Article 27 Decisions, authorisations, authorisation renewals, technical amendments, authorisations closed etc.
2. Refers to legislation to be published so therefore no entry required for the EPA.
3. Datasets published to DPER Open Portal from 2015. 2013 and 2014 figures are datasets published to EPA website.
4. Includes visits to IPC, IE, Waste, Dumping at Sea and VOC permitted sites.
Source: Information provided by EPA.

The EPA prepares an Annual Report and Accounts that is sent to the Minister of DCCAE, who lays it before both Houses of Oireachtas. Once the Annual Report has been approved by the Oireachtas, the EPA publishes the report online (EPA, 2017[2]). The report is structured in line with four out of five of the goals of the EPA's strategic plan (regulation, knowledge, advocacy, and "organisationally excellent", omitting "responding to key environmental challenges").

There is no structured mechanism of engagement around the EPA's performance reporting with the legislative branch, but the EPA is often called to appear before Joint Oireachtas Committees to discuss particular issues, or to submit written answers to Parliamentary Questions. The EPA can also be invited into the Public Accounts Committee, which has happened on three occasions since its founding.

Notes

[1] Section 55 states that "The Agency may, of its own volition, and shall when requested by a Minister of the Government, give information or advice or make recommendations for the purposes of environmental protection to any such Minister on any matter relating to his functions or responsibilities and that Minister shall have regard to any such information or advice given or recommendations made".

[2] In general, for amounts under EUR 5 000 EUR the EPA can directly procure goods and services; for amounts between EUR 5 000 EUR and EUR 25 000 EUR the EPA can use a restricted procurement procedure; for amounts above EUR 25 000 the EPA has to carry out an open tender on the central government website with the process taking up to two months; and for amounts above the EU threshold of EUR 209 000 EUR the EPA has to carry out an open tender for a longer period with the whole process taking 4-6 months. The EPA can use an accelerated procurement process for urgent requirements e.g. services that may be required for safety cases.

References

DPER (2016), *Code of Practice for the Governance of State Bodies*, Department of Public Expenditure and Reform, Ireland, https://govacc.per.gov.ie/wp-content/uploads/Code-of-Practice-for-the-Governance-of-State-Bodies.pdf. [4]

EPA (2019), *EPA Educational Resources*, http://www.epa.ie/researchandeducation/education/ (accessed on 12 September 2019). [17]

EPA (2019), *Guidance on Requests for Alterations to a Licensed Industrial or Waste Activity*, https://www.epa.ie/pubs/advice/licensee/Licence%20Alteration%20Guidance%20rev%20MOC%2021-06-19.pdf (accessed on 16 September 2019). [10]

EPA (2019), *Making an Environmental Complaint - Environmental Protection Agency (EPA*, Environmental Protection Agency (EPA), http://www.epa.ie/enforcement/report/ (accessed on 12 September 2019). [16]

EPA (2019), *National Waste Prevention Programme*, http://www.epa.ie/waste/nwpp/ (accessed on 12 September 2019). [18]

EPA (2019), *Our Commitment To Customer Service*, Environmental Protection Agency (EPA), http://epa.ie/about/qcs/ (accessed on 12 September 2019). [15]

EPA (2018), *Drinking Water Report for Public Supplies 2017*, http://www.epa.ie/pubs/reports/water/drinking/drinkingwaterreport2017.html (accessed on 12 September 2019). [20]

EPA (2018), *SEA Pack*, Environmental Protection Agency (EPA). [3]

EPA (2018), *Strategic Plan2016-2020: Our Environment Our Wellbeing*, http://www.epa.ie/pubs/reports/other/corporate/EPA_StrategicPlanWeb_2018.pdf. [5]

EPA (2017), *Environmental Protection Agency Annual Report and Accounts 2017*, http://www.epa.ie/pubs/reports/other/corporate/EPA_AnnualReport_2017_EN_web.pdf. [2]

EPA (2017), *EPA Industrial and Waste Licence Enforcement*, https://www.epa.ie/pubs/reports/enforcement/EPA_Industrial_Waste_LE_Report2017.pdf. [11]

EPA (2017), *Urban Waste Water Treatment in 2016*, http://www.epa.ie/pubs/reports/water/wastewater/uwwreport2016.html (accessed on 12 September 2019). [21]

EPA (2016), *Draft Guidance on Article 15 of Industrial Emissions Directive (2010/75/EU)*, http://epa.ie/pubs/advice/licensee/Draft%20Guidance%20on%20IED%20alternate%20ELVs%20or%20derogation%20from%20BAT%20AELs.pdf (accessed on 16 September 2019). [14]

EPA (2016), *EPA Industrial and Waste Licensing Enforcement*, https://www.epa.ie/pubs/reports/enforcement/EPA_industrial_waste_licence_enforcementReport2016.pdf. [12]

EPA (2015), *2015 EPA Industrial and Waste Licence Enforcement*, http://www.epa.ie/pubs/reports/enforcement/EPAIndustrialandWasteLicenceEnforcement2015.pdf. [13]

EPA Review Group (2011), *A Review of the Environmental Protection Agency*, https://www.epa.ie/pubs/reports/other/corporate/EPA%20Review%20Report.pdf (accessed on 16 September 2019). [8]

European Commission (2019), *The Aarhus Convention*, https://ec.europa.eu/environment/aarhus/index.htm. [19]

IPA (2018), "A Case Study of the EPA-RPII Merger", *State of the Public Service Series* Research Paper 22, https://ipa.ie/_fileUpload/Documents/EPA_RESEARCHEPORT_2018.pdf (accessed on 16 September 2019). [9]

McDonagh, J., L. Burke and G. O'Leary (2018), "Tracing the evolution and development of a senior management network in a government organisation: The case of the Environmental Protection Agency", *Administration*, Vol. 66/4, pp. 27-48, http://dx.doi.org/10.2478/admin-2018-0031. [7]

OECD (2017), *Creating a Culture of Independence: Practical Guidance against Undue Influence*, The Governance of Regulators, OECD Publishing, Paris, https://dx.doi.org/10.1787/9789264274198-en. [6]

Shipan, C. (2006), "Independence and the Irish Environmental Protection Agency: A Comparative Assessment", *Studies in Public Policy*, Vol. 20, pp. 1-82. [1]

Annex A. Methodology

Measuring regulatory performance is challenging, starting with defining what to measure, dealing with confounding factors, attributing outcomes to interventions and coping with the lack of data and information. This chapter describes the methodology developed by the OECD to help regulators address these challenges through a Performance Assessment Framework for Economic Regulators (PAFER), which informs this review. The chapter first presents some of the work conducted by the OECD on measuring regulatory performance. It then describes the key features of the PAFER and presents a typology of performance indicators to measure input, process, output and outcome. It finally provides an overview of the approach and practical steps undertaken for developing this review.

Analytical framework

The analytical framework that informs this review draws on the work conducted by the OECD on measuring regulatory performance and the governance of economic regulators. OECD countries and regulators have recognised the need for measuring regulatory performance. Information on regulatory performance is necessary to better target scarce resources and to improve the overall performance of regulatory policies and regulators. However, measuring regulatory performance can prove challenging. Some of these challenges include:

- *What to measure*: evaluation systems require an assessment of how inputs have influenced outputs and outcomes. In the case of regulatory policy, the inputs can focus on: i) overall programmes intended to promote a systemic improvement of regulatory quality; ii) the application of specific practices intended to improve regulation, or, iii) changes in the design of specific regulations.
- *Confounding factors*: there is a myriad of contingent issues that have an impact on the outcomes in society which regulation is intended to affect. These issues can be as simple as a change in the weather, or as complicated as the last financial crisis. Accordingly, it is difficult to establish a direct causal relationship between the adoption of better regulation practices and specific improvements to the welfare outcomes that are sought in the economy.
- *Lack of data and information*: countries tend to lack data and methodologies to identify whether regulatory practices are being undertaken correctly and what impact these practices may be having on the real economy.

The OECD (2014[1]) *Framework for Regulatory Policy Evaluation* starts addressing these challenges through an input-process-output-outcome logic, which breaks down the regulatory process into a sequence of discrete steps. The input-process-output-outcome logic is flexible and can be applied both to evaluate practices to improve regulatory policy in general, and also to evaluate regulatory policy in specific sectors, based on the identification of relevant strategic objectives. It can be tailored to economic regulators by taking into consideration the conditions that support the performance of economic regulators (Box A A.1).

The OECD Best Practice Principles for Regulatory Policy: The Governance of Regulators (OECD, 2014[2]) identifies some of the conditions that support the performance of economic regulators. They recognise the importance of assessing how a regulator is directed, controlled, resourced and held to account, in order to improve the overall effectiveness of regulators and promote growth and investment, including by supporting competition. Moreover, they acknowledge the positive impact of the regulator's own internal process on outcomes (i.e. how the regulator manages resources and what processes the regulator puts in place to regulate a given sector or market) (Figure A A.1).

> **Box A A.1. The input-process-output-outcome logic sequence**
>
> - Step I. Input: indicators include for example the budget and staff of the regulatory oversight body.
> - Step II. Process: indicators assess whether formal requirements for good regulatory practices are in place. This includes requirements for objective setting, consultation, evidence-based analysis, administrative simplification, risk assessments and aligning regulatory changes internationally.
> - Step III. Output: indicators provide information on whether the good regulatory practices have actually been implemented.

> - Step IV. Impact of design on outcome (also referred to as intermediate outcome): indicators assess whether good regulatory practices contributed to an improvement in the quality of regulations. It therefore attempts to make a causal link between the design of regulatory policy and outcomes.
> - Step V. Strategic outcomes: indicators assess whether the desired outcomes of regulatory policy have been achieved, both in terms of regulatory quality and in terms of regulatory outcomes.
>
> Source: (OECD, 2014[1]).

Figure A A.1. The OECD Best Practice Principles on the Governance of Regulators

Source: Adapted from (OECD, 2014[2]).

The two frameworks are brought together into a Performance Assessment Framework for Economic Regulators that structures the drivers of performance along the input-process-output-outcome framework (Table A A.1).

Table A A.1. Criteria for assessing regulators' own performance framework

References	Strategic objectives	Input	Process	Output and outcome
Best Practice Principles for the Governance of Regulators	• Role clarity	• Funding	• Maintaining trust and preventing undue influence	• Performance evaluation
			• Decision making and governing body structure	
			• Accountability and transparency	
			• Engagement	
Institutional, organisational and monitoring drivers?	• Objectives and targets	• Budgeting and financial management	• Strategy, leadership and co-ordination	• Performance standards and indicators
	• Functions and powers	• Human resources management	• Institutional structure	• Performance processes and reports
			• Management systems and operating processes	• Feedback or outside evidence on performance
			• Relations and interfaces with Government bodies, regulated entities and other key stakeholders	
			• Regulatory management tools	

Source: OECD Analysis.

Performance indicators

For regulators, performance indicators need to fit the purpose of performance assessment, which is a systematic, analytical evaluation of the regulator's activities, with the purpose of seeking reliability and usability of the regulator's activities. Performance assessment is neither an audit, which judges how employees and managers complete their mission, nor a control, which puts emphasis on compliance with standards (OECD, 2004[3]).

Accordingly, performance indicators need to assess the efficient and effective use of a regulator's inputs, the quality of regulatory processes, and identify outputs and some direct outcomes that can be attributed to the regulator's interventions. Wider outcomes should serve as a "watchtower", which provides the information the regulator can use to identify problem areas, orient decisions and identify priorities (Figure A A.2).

Figure A A.2. Input-process-output-outcome framework for performance indicators

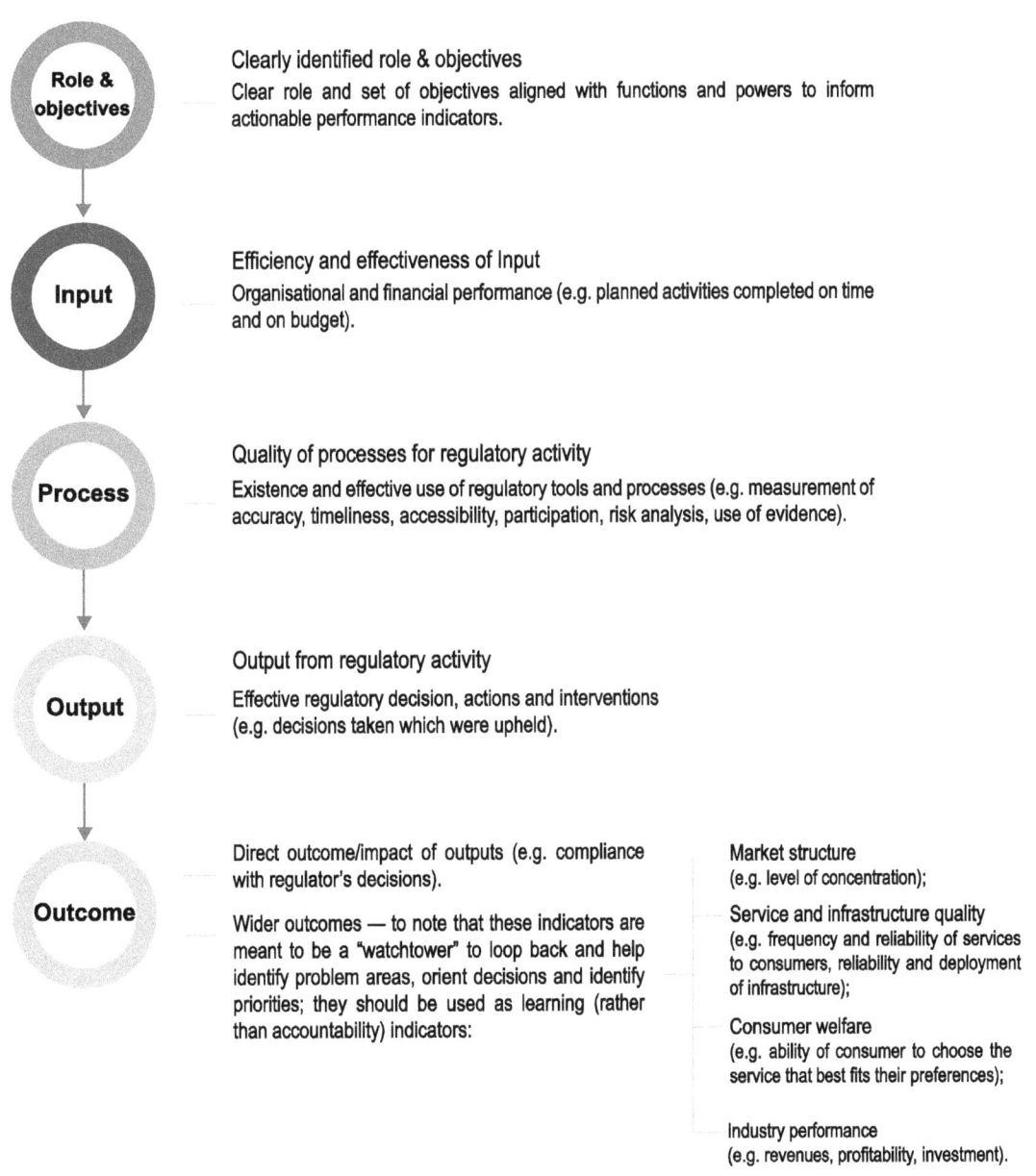

Notes: This framework was proposed in the initial methodology for the performance assessment framework for economic regulators (PAFER) discussed with the OECD Network of Economic Regulators (NER). It has been refined to reflect feedback from NER members and the experience of other regulators in assessing their own performance.
Source: (OECD, 2015[4]), Figure 3.3 (updated in 2017).

Approach

The analytical framework presented above informed the data collection and the analysis presented in the report. The present report looks at the internal and external governance arrangements of Ireland's Environmental Protection Agency (EPA) in the following areas:

- **Strategic objectives**: to identify the existence of a set of clearly identified objectives, targets, or goals that are aligned with the regulator's functions and powers, which can inform the development of actionable performance indicators;
- **Input**: to determine the extent to which the regulator's funding and staffing are aligned with the regulator's objectives, targets or goals, and the regulator's ability to manage financial and human resources autonomously and effectively;
- **Process**: to assess the extent to which processes and the organisational management support the regulator's performance;
- **Output and outcome**: to identify the existence of a systematic assessment of the performance of the regulated entities, the impact of the regulator's decisions and activities, and the extent to which these measurements are used appropriately.

Data informing the analysis presented in the report was collected via a desk review, two fact-finding missions and a peer mission to Ireland:

- **Questionnaire and desk review**: EPA completed a detailed questionnaire which informed a desk review by the OECD Secretariat. The Secretariat reviewed existing legislation and EPA documents to collect information on the *de jure* functioning of the regulator, and to inform the basis of the fact-finding missions. This questionnaire was tailored to EPA, based on the methodology already applied by the OECD to Colombia's Communications Regulation commission (OECD, 2015[4]), Latvia's Public Utilities Commission (OECD, 2016[5]), Mexico's three energy regulators (OECD, 2017[6]); (OECD, 2017[7]); (OECD, 2017[8]); (OECD, 2017[9]), Ireland's Commission for Regulation of Utilities (OECD, 2018[10]); Peru's Energy and Mining Regulator (OECD, 2019[11]); Peru's Telecommunications Regulator (OECD, 2019[12]).
- **Fact-finding missions**: the first fact-finding mission focused primarily on internal governance and was conducted by the OECD Secretariat on 18-21 February 2019 at three EPA offices: Wexford (HQ), Cork and Kilkenny. The second fact-finding mission took place in Dublin on 12-14 March 2019 and focused primarily on external governance. These missions were the key tool to collect and complete the *de jure* information obtained through the questionnaire with the *de facto* state of play. The work of the fact-finding missions tailored the PAFER methodology to EPA features. Information collected was completed and checked with EPA for accuracy and issues for further discussion were also flagged.
- **Peer mission**: the mission took place on 18-21 June 2019 in Dublin and included peer reviewers from Norway and the United Kingdom (Scotland), in addition to OECD Secretariat. This mission met with key stakeholders in EPA as well as externally. At the end of the mission, the team discussed preliminary findings and recommendations jointly with senior management from EPA to test their feasibility and goodness of fit.

During the fact-finding and peer missions, the team met with EPA's leadership team as well as a number of staff from across the institution. In addition, the team met with government institutions and external stakeholders, including:

- Department of Communications, Climate Action and Environment
- Department of Housing, Planning and Local Government
- Department of Agriculture, Food and the Marine
- Joint Oireachtas Committee on Communications, Climate Action and Environment
- Members of the EPA Advisory Committee
- National Economic and Social Council
- Climate Change Advisory Council
- Commission for the Regulation of Utilities
- Health and Safety Authority
- Planning Appeals Board
- Kilkenny County Council
- Laois County Council
- Dublin County Council
- Local Authority Environment Committee
- Ireland Environment Network
- Irish Farmers Association
- Irish Business and Employers Confederation
- Irish Waste Management Association
- Pfizer Group
- University College Cork Environmental Research Institute

The Secretariat also held phone interviews with:

- Health Services Executive
- DG Environment, European Commission
- European Environment Agency

References

OECD (2019), *Driving Performance at Peru's Energy and Mining Regulator*, The Governance of Regulators, OECD Publishing, Paris, https://dx.doi.org/10.1787/9789264310865-en. [11]

OECD (2019), *Driving Performance at Peru's Telecommunications Regulator*, The Governance of Regulators, OECD Publishing, Paris, https://dx.doi.org/10.1787/9789264310506-en. [12]

OECD (2018), *Driving Performance at Ireland's Commission for Regulation of Utilities*, The Governance of Regulators, OECD Publishing, Paris, http://dx.doi.org/10.1787/9789264190061-en. [10]

OECD (2017), *Driving Performance at Mexico's Agency for Safety, Energy and Environment*, The Governance of Regulators, OECD Publishing, Paris, https://dx.doi.org/10.1787/9789264280458-en. [9]

OECD (2017), *Driving Performance at Mexico's Energy Regulatory Commission*, The Governance of Regulators, OECD Publishing, Paris, https://dx.doi.org/10.1787/9789264280830-en. [7]

OECD (2017), *Driving Performance at Mexico's National Hydrocarbons Commission*, The Governance of Regulators, OECD Publishing, Paris, https://dx.doi.org/10.1787/9789264280748-en. [8]

OECD (2017), *Driving Performance of Mexico's Energy Regulators*, The Governance of Regulators, OECD Publishing, Paris, https://dx.doi.org/10.1787/9789264267848-en. [6]

OECD (2016), *Driving Performance at Latvia's Public Utilities Commission*, The Governance of Regulators, OECD Publishing, Paris, https://dx.doi.org/10.1787/9789264257962-en. [5]

OECD (2015), *Driving Performance at Colombia's Communications Regulator*, OECD Publishing, Paris, https://dx.doi.org/10.1787/9789264232945-en. [4]

OECD (2014), *OECD Framework for Regulatory Policy Evaluation*, OECD Publishing, Paris, https://dx.doi.org/10.1787/9789264214453-en. [1]

OECD (2014), *The Governance of Regulators*, OECD Best Practice Principles for Regulatory Policy, OECD Publishing, Paris, https://dx.doi.org/10.1787/9789264209015-en. [2]

OECD (2004), *The choice of tools for enhancing policy impact: Evaluation and review*, OECD, Paris, http://www.oecd.org/officialdocuments/publicdisplaydocumentpdf/?cote=gov/pgc(2004)4&doclanguage=en (accessed on 16 November 2018). [3]

www.ingramcontent.com/pod-product-compliance
Ingram Content Group UK Ltd.
Pitfield, Milton Keynes, MK11 3LW, UK
UKHW050413240426
12048UKWH00020B/1486